HOW THE
World
LOOKS TO A
Bee

D1502825

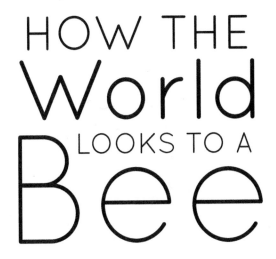

HOW THE World LOOKS TO A Bee

AND OTHER MOMENTS OF SCIENCE

Edited by

Don Glass

INDIANA UNIVERSITY PRESS

This book is a publication of

Indiana University Press
Office of Scholarly Publishing
Herman B Wells Library 350
1320 East 10th Street
Bloomington, Indiana 47405 USA
iupress.indiana.edu

© 2020 by WFIU/RTVS

All rights reserved

No part of this book may be reproduced or utilized in any
form or by any means, electronic or mechanical, including
photocopying and recording, or by any information storage
and retrieval system, without permission in writing from
the publisher. The paper used in this publication meets
the minimum requirements of the American National
Standard for Information Sciences—Permanence of Paper
for Printed Library Materials, ANSI Z39.48–1992.

Manufactured in the United States of America

Library of Congress Cataloging-in-Publication Data

Names: Glass, Don [date] | WFIU (Radio
    station: Bloomington, Ind.)
Title: How the world looks to a bee: and other
    moments of science / [edited by] Don Glass.
Other titles: Moments of science
Description: Bloomington, Indiana: Indiana University
    Press, [2020] | Based on radio scripts from program A
    moment of science, on public radio station WFIU-FM. |
    Includes bibliographical references and index.
Identifiers: LCCN 2019011402 (print) | LCCN 2019016209
    (ebook) | ISBN 9780253046284 (ebook) | ISBN 9780253046253
    (hardback: alk. paper) | ISBN 9780253046260 (pbk.: alk. paper)
Subjects: LCSH: Science—Popular works. |
    Moment of science (Radio program)
Classification: LCC Q162 (ebook) | LCC Q162
    .H7925 2020 (print) | DDC 500—dc23
LC record available at https://lccn.loc.gov/2019011402

1 2 3 4 5   25 24 23 22 21 20

Contents

Acknowledgments

The chapters in this book are based on scripts written for the WFIU radio series *A Moment of Science* by Barbara Bolz, Rory Boothe, Amy Breau, Danit Brown, Stephen Fentress, Luca Fitzgerald, Don Glass, Susan Linville, Heather Love, Sara Loy, Victoria Miluch, William Orem, Paul Patton, Michelle Ross, Jeremy Shere, Eric Sonstroem, and Don Ulin.

HOW THE World LOOKS TO A Bee

Does NutraSweet Have Calories?

Picture this: One day, a coworker unexpectedly showered you with compliments—your tie, your hair, your shirt. She was normally quite friendly, but this was over the top. You asked her why she was being so sweet, and she explained that she had been reading about artificial sweeteners and had decided to be artificially sweet that day. You hinted that she was laying it on a bit thick, and she pointed out that that's what artificial sweeteners do as well.

She elaborated by explaining that saccharin and sucralose, which is also known as Splenda, are hundreds of time sweeter than sugar. And aspartame, also known as NutraSweet, is about 160 times sweeter. That means that one teaspoon of aspartame is the same as 160 teaspoons of sugar. This sweetness is why you can add aspartame to your food without adding calories.

Not to be outdone in the know-it-all department, you suggested she might be confusing two issues. The first was how sweet a substance tastes, which has to do with how well the molecules that make up this substance chemically bind with the sweet taste receptors in our mouth. The second was the amount of energy released when we metabolize, or digest, this substance, which is measured in calories. So the reason saccharin and sucralose have no calories is because our bodies don't metabolize them.

She replied that in fact we do metabolize aspartame, and it breaks down into chemicals that have a caloric value.

However, you thought you had her when you replied that diet soda is sweetened with aspartame and has no calories.

But it wasn't over yet, because she replied by showing that's where the degree of sweetness comes in. Because aspartame is 160 times as sweet as sugar, you only need to use a fraction of the amount of sugar you'd have to use otherwise. This amount is so small that it's insignificant in terms of calories.

You had to agree that was, well . . . sweet.

Further Reading

Purves, William K. "How Can an Artificial Sweetener Contain No Calories?" *Scientific American.* Accessed June 4, 2019. https://www.scientificamerican.com/article/how-can-an -artificial-swe/.

WebMD. "Stevia and Sugar Substitutes." Last reviewed February 16, 2019. https://www.webmd.com/diet/stevia-sugar -substitutes#1.

A Water Magnifier

Punch a hole about an eighth of an inch in diameter in the bottom of a paper or foam cup. Now push the cup down into a deep bowl of water. Look down into the cup while you push. Of course you'll see water come in through the hole you punched. But as the cup fills, you'll notice something else. You'll notice that the hole in the bottom of the cup appears magnified. You can make the hole appear bigger by pushing down harder on the cup. Push down more gently, and the magnification is reduced.

This happens because rays of light are bent when they cross from one medium to another. In this case, rays of light that make up the image of the hole in the bottom of the cup are bent as they cross from water into air. The rays are bent in just the right way to create a magnified image of the hole from your point of view.

The reason those light rays bend in that particular way has to do with the shape of the water surface. When you push down on the cup, water spurts upward through the hole in the bottom. That upward-spurting stream of water makes a bulge in the water surface more or less like the bulge in the surface of a glass magnifying lens. By varying the downward pressure on the cup, you can vary the size of the bulge in the water surface. That, in turn, varies the amount of magnification.

Further Reading

McMath, T. A. "Refraction—A Surface Effect." *Physics Teacher* 27, no. 3 (March 1989). https://doi.org/10.1119/1.2342712.

Conversation at a Crowded Party

"I don't know what she sees in him."

"Beg your pardon?"

"I said, I DON'T KNOW WHAT SHE SEES IN HIM."

An article published in 1959 in the *Journal of the Acoustical Society of America* presented a rough theoretical analysis of sound levels at cocktail parties. The author, William MacLean, analyzed the problem of carrying on a conversation in the presence of background noise from other conversations, and he made a prediction you can check for yourself.

At the beginning of a party, when few guests are present, quiet conversation is possible. As more guests arrive, you have to talk louder and louder to override the increasing background noise.

MacLean's calculations predicted that when the number of guests at a party exceeds a certain maximum determined by the size and other characteristics of the room, merely talking louder is of no avail in continuing your conversation. You just force everybody else to talk louder. The ensuing increase in background noise soon drowns you out unless you move closer to the person you're talking to—closer than you might get in another situation. The acoustics of real rooms are so complex that it's practically impossible to say exactly when this need to get closer will set in, but MacLean predicted the moment will occur at some point as more and more people arrive.

Someone may temporarily quiet a loud party, perhaps to introduce the guest of honor. But, MacLean found, even if everyone tries to talk quietly afterward, dialogues like the one we began with eventually drive the background noise back up to its earlier level. A crowded party remains loud until guests begin to leave.

Further Reading

Hall, Edward T., Jr. "The Anthropology of Manners." *Scientific American* 192, no. 4 (April 1955): 84–91.

MacLean, William R. "On the Acoustics of Cocktail Parties." *Journal of the Acoustical Society of America* 31, no. 1 (January 1959). https://doi.org/10.1121/1.1907616.

Can a Theory Evolve into a Law?

In this Moment of Science we clear up the difference between a scientific theory and a scientific law. You know, it's something like the argument that if the theory of evolution were true, it would actually be a law. In fact, scientists get a little weary of some people saying the fact that evolution is a theory means modern science itself isn't convinced it really happens.

So it seemed like it would be a good idea to go over the terminology one more time.

Well, the definition of a law is easy. It's a description—usually mathematical—of some aspect of the natural world, such as gravity. The law of gravity describes and quantifies the attraction between two objects. But the law of gravity *doesn't* explain what gravity is or why it might work in this way. That's because that kind of explanation falls into the realm of theory.

And the theory that *explains* gravity is the theory of general relativity.

According to the National Academy of Sciences, a scientific theory is a "well-substantiated explanation of some aspect of the natural world that can incorporate facts, laws, inferences, and tested hypotheses." In other words, all scientific theories are supported by evidence, and you can test them, and—most important—you can use them to make predictions.

So based on that definition, theories never change into laws, no matter how much evidence out there supports them. Formulating theories, in fact, is the end goal of science. So to say evolution is just a theory is actually an argument for it and not against it. You can't do any better in science than to be a theory.

Further Reading

Patheos. "Evolution Is Just a Theory!" Accessed June 4, 2019. https://www.patheos.com/blogs/daylightatheism/essays /evolution-is-just-a-theory.

Rennie, John. "15 Answers to Creationist Nonsense." *Scientific American*, July 1, 2002. https://www.scientificamerican .com/article/15-answers-to-creationist/.

A Cat "Flips Out"

If you were asked to guess why cats flip toys up into the air, what would you answer? If you answered that the cat was pretending the toy was a bird, that would be a good guess because they certainly look like they're pretending to catch birds. How could we test this idea?

Well, how about studying cats actually catching birds and seeing if the motions are the same? If you did this, you'd find that cats have three basic attack strategies. One is "stalk, pounce, grab with front claws, and bite." Let's call that the "mouse" pattern. Another is "stalk, leap up with front paws swinging, and bite." That's the actual "bird" pattern, and it works for prey that may fly off during the attack.

The third may surprise you. Let's call it the "fish" pattern. You might raise your eyebrows at the suggestion that cats can fish, but many cats do catch fish out of shallow rivers, just like bears. Guess what the fish pattern looks like? Do you think they flip it into the air? In fact, they reach under the fish and quickly flip it backward over their shoulders, then turn and pounce. It's the same pattern a playing cat shows when flipping a toy in the air! Many folks have seen a cat catch a mouse, but not a fish, which is probably why we don't immediately realize what they're practicing when they play.

Further Reading

Morris, Desmond. *Catwatching*. New York: Crown Publishers, 1987.

Winter Sounds

For those of us who live in the northern part of the country, arguably no other season transforms the environment quite as much as winter. Snow makes the world over so that it's hardly recognizable.

The scenery isn't the only thing that changes during the winter season, though. Winter also transforms the way the environment sounds.

Before that first snowfall, when the ground has hardened from the cold temperatures, sound experiences its first winter makeover. Just as smooth, hard surfaces such as glass reflect light, smoother and harder surfaces better reflect sound. The frozen surfaces of the early winter season absorb less sound. Thus, acoustic waves retain more of their energy when they reflect off the ground. As a result, early winter sounds are louder and clearer. Pay attention, and you might notice that everything sounds a little bit crisper.

After the first snow, sound is made over again. The porousness of snow makes it absorb sound. Specifically, snow tends to absorb higher-frequency acoustical waves. As a result, sound becomes quieter and somewhat distorted. Without looking out your window, you might know it had snowed simply from noticing how bird calls or passing cars are slightly muffled.

If the snow lingers, or it is topped with rain, winter sound will be made over once more. As snow hardens, it becomes smoother, less porous. As you can probably guess, this means the snow will, like the hardened ground that preceded it, reflect sound well. Sound will once again be crisp and clear.

This winter, see if you can hear how the season transforms sound as well as the scenery.

Further Reading

National Snow and Ice Data Center. "Snow Characteristics." Accessed June 4, 2019. https://nsidc.org/cryosphere/snow /science/characteristics.html.

Wikipedia. "Sound." Accessed July 6, 2019. http://en.wikipedia .org/wiki/Sound_wave.

Rust

Water and oxygen are both necessary for metal to rust, but equally important is the flow of electrons from the metal into the surrounding water. In order for iron to be dissolved in water, it has to have some kind of electrical charge. All atoms begin with a balanced number of negatively charged electrons and positively charged protons. So when the iron gives up electrons, it then has a net positive charge.

Once the iron is positively charged, it can react with oxygen dissolved in the water to form iron oxide, or rust. As the rust forms, the iron loses its positive charge. Now it can give up more electrons to the water, and so the cycle continues.

Usually rust forms at two different sites in two different ways. For example, a spike stuck partway into the ground rusts fastest underground, where the metal can give up electrons into the moist soil. As the spike gives up electrons, the whole spike becomes positively charged. When that happens, the metal at the top of the spike, which is exposed to the oxygen in the air, also forms a coating of rust.

We've seen how metal exposed to salt, whether from seawater or salt put on the roads in winter to melt snow and ice, rusts faster. Salt speeds up corrosion because salt water conducts electricity better than fresh water, making it easier for iron to give up electrons.

In the 1930s, scientist Michael Faraday was the first to explain metal corrosion as an electrochemical process. Since then we have learned a lot more, but there is still a great deal that is not known about how rust works. A more recent wrinkle is that scientists have found that bacteria, too, are involved in metal corrosion, making it a biological process as well. So if you drive an old car, it may have even more bugs in it than you think. Rust involving bacteria is called "biocorrosion."

Bacteria can work to either slow down or speed up rust by creating what is called a "biofilm" over the metal. This biofilm, made up of living bacteria, can act as a coat of paint or rustproofing, protecting the metal from the corrosive effects of moisture. But the biofilm can also have the opposite effect if other types of bacteria get to the metal first. For example, sulfate-reducing bacteria draw dissolved sulfur out of the water and release sulfuric acid. Sulfuric acid is extremely attractive to electrons and causes the metal to corrode much faster than normal. But sulfate-reducing bacteria work better when there is no oxygen present. If the sulfate-reducing bacteria get to the metal before the biofilm forms, the film can actually cover and protect the damaging bacteria from exposure to oxygen and water currents. Sealed under the biofilm, the sulfate-reducing bacteria can work away at the metal relatively undisturbed.

Further Reading

Collier's Encyclopedia, s.v. "Metals, Corrosion of." New York: Macmillan Educational Company, 1987.

Sienko, Michell, and Robert A. Plane. *Chemistry*. 5th ed. New York: McGraw-Hill, 1976.

Hungry Lasagna

So you made lasagna last night, and it was delicious. You saved the leftovers by wrapping aluminum foil over the top of the pan and putting it in the fridge. When you get hungry today, you go to look for that lasagna, but it looks like the lasagna was getting hungry, too! It has eaten tiny holes in the aluminum foil that was covering it! Why does this happen?

There are really two things responsible for those holes in the aluminum foil: the acidic nature of the lasagna and some curious properties of aluminum. Lasagna gets most of its acid from the tomatoes in its sauce. Almost everything you eat is at least slightly acidic, but tomatoes are especially so. This acid wouldn't be a problem for a glass container, or stainless steel, or plastic wrap, or for most materials we use to store and prepare food. Aluminum, though, is especially vulnerable to acid.

Here's why: Most metals form a protective layer on their surface called an *oxide layer*. Aluminum forms an oxide layer, too, but it is very thin, allowing the tomato's acid to easily break through. This lets the tomato sauce dissolve the aluminum, and it creates gray or black chemicals on top of the lasagna that taste very bad.

The same thing can happen if you use aluminum cookware for acidic sauces. Sauce made in an aluminum pot will be grayer and not as tasty as sauce made in other cookware.

Further Reading

McGee, Harold. *On Food and Cooking: The Science and Lore of the Kitchen*. New York: Scribner, 1984.

Centoni, Danielle. "Turns Out Aluminum Foil Isn't as Simple as You Thought." *OregonLive*, April 27, 2010. https://www.oregonlive.com/foodday/2010/04/take_a_shine_to_aluminum_foil.html.

A Wet Paintbrush

Take an artist's watercolor brush, dip it in water, and pull it out again. The bristles cling together to form a smooth, pointed shape that artists and calligraphers use to paint lines of varying thickness. Something similar happens when a person with straight hair dives into water and climbs out again: that person's hair is slicked down.

We usually say the bristles or hairs cling together because they are wet. But that can't be the real explanation, as you can see if you look at the bristles of that brush while you hold it underwater. While immersed in water, the bristles do not cling, even though they are certainly wet. Wet bristles—and wet human hairs—cling together not if they are surrounded by water, but only if they are surrounded by a water surface.

The clinging of the bristles is really a manifestation of the clinging of individual water molecules. A water molecule at the surface of a body of water—say, on the outside of a wet brush—is pulled strongly toward that body of water because that's where the other water molecules are. One result of this mutual attraction between water molecules is that a water surface is under tension, like an elastic skin.

That surface tension pulls water into beads on a well-waxed car. It also holds the bristles of a wet brush together—if the brush is surrounded by air. The bristles of a brush immersed in water don't cling because they are not surrounded by a water surface.

Further Reading

Boys, C. V. *Soap Bubbles and the Forces Which Mould Them*. Garden City, NY: Doubleday Anchor Books, 1959.

The Glory

Next time you fly, try to get a window seat on the shady side of the plane, not above the wing. You want a clear view of the airplane's shadow on the clouds below. Look for a bull's-eye pattern of bright rings, tinged with rainbow color, surrounding the shadow on the clouds: the so-called glory.

The glory is not the same as a rainbow. A rainbow is made of sunlight coming back from water droplets at an angle of about forty-two degrees from the direction it went in. The glory, on the other hand, is light coming out of the droplet in a direction almost exactly opposite the incoming sunlight.

Physicists have found the explanation extremely complicated, but part of the story goes like this: Light encountering a spherical water droplet in a cloud has, so to speak, several options. One is to enter one side of the water droplet at a glancing angle, bounce once off the inner surface of the back of the droplet, and then exit the droplet on the other side—basically making a U-turn. Another option for a light ray is to bounce not once but fourteen times off the inside of the droplet, making three and a half trips around the droplet in the process (like a car searching a lot for a parking space), and then exit the droplet.

It turns out that light that has bounced once and light that has bounced fourteen times will emerge from the droplet going in almost exactly the same direction. A doubly strong light ray goes almost straight back toward the sun and, hence, toward your airplane. But this complicated process also breaks white light into its colors, and different colors emerge in slightly different directions. That's why the glory is tinged with color.

So, if you have a window seat on the shady side of a plane flying not too high over clouds, look for the glory: a mysterious bull's-eye pattern of colored rings surrounding the shadow of the airplane.

Bibliography

Bryant, Howard C., and Nelson Jarmie. "The Glory." *Scientific American* 231, no. 1 (July 1974): 60–73. Reprinted in *Atmospheric Phenomena*, edited and with an introduction by David K. Lynch. New York: W. H. Freeman, 1980.

Horns versus Antlers

Have you ever wondered what the difference is between horns and antlers? In fact, horns and antlers have some important things in common. They both grow from an animal's forehead and are used to assert dominance, provide defense, and attract mates. Although it can be easy to mix them up, there are some simple ways to remember which is which.

First of all, looks are important. Horns look like daggers, sometimes twisted into exotic shapes. Antlers look like branches, with multiple points.

Second, different types of animals sport different types of headgear. Horns belong to the bovids—animals such as sheep, goats, cows, and bison. Antlers belong to the cervids. That includes all deer, elk, moose, and caribou (or reindeer).

Number three: Horns are made of keratin just like your fingernails, which means they grow throughout an animal's whole life and never fall off. Antlers, on the other hand, are made of bone. They fall off every year in winter or early spring, and a whole new set grows in time for mating season in the fall.

Which brings us to our last big difference. Horns, you see, usually grow on both the male and female members of a species, while antlers belong almost exclusively to the boys. Female caribou are the only exception, and their antlers are pretty small.

So, here's a little word exercise you can do that might help you remember:

Horns	Antlers
Daggers	Branches
Bovids	Cervids
Keratin	Bone
Both sexes	Just the guys

Further Reading

Lincoln, G. A. "Biology of Antlers." *Journal of Zoology* 226, no. 3 (March 1992): 517–28.

Myers, P., R. Espinosa, C. S. Parr, T. Jones, G. S. Hammond, and T. A. Dewey. "Horns and Antlers." Animal Diversity Web. 2019. https://animaldiversity.org/collections /mammal_anatomy/horns_and_antlers/.

National Park Service: Yellowstone National Park. "Difference between Antlers and Horns." Accessed March 25, 2019. http://www.nps.gov/yell/forkids/ahdiff.htm.

Glacier "Sawdust"

The Colorful Component of Mountain Lakes

Have you ever stood on the shore of an icy-cold mountain lake and marveled at its dazzling shade of blue? If so, you're in luck, because this Moment of Science is all about the role glaciers play in giving those chilly lakes such incredible turquoise colors.

Here's the scoop: Glaciers work kind of like gigantic sheets of sandpaper. You know how you create a lot of fine sawdust when you rub the coarse side of sandpaper over a wooden surface? Well, glaciers do pretty much the same thing, except they work on mountain rock rather than wood.

Glacial ice constantly moves very slowly downhill. While it moves, bits of rock and gravel get stuck between the ice and the mountain, forming a coarse surface similar to the sand on sandpaper. The rocks grind together, creating the glacier equivalent of sawdust. Geologists call this dust "rock flour," and as glaciers melt during spring and summer, they transport it down into mountain lakes.

Here's where things get colorful. The rock flour is so fine that it doesn't sink to the bottom of the lake. Instead, it remains suspended throughout the water. When sunlight hits, the water absorbs the long-wave colors of the spectrum: the reds, oranges, and yellows. At the same time, the rock flour absorbs some of the shortest light waves—the purples and indigos—and scatters the remaining light back to our eyes. Thanks to these processes, the reflected light is mostly green, with a dash of blue. And voilà— turquoise water!

The next time you come across one of these dazzling lakes, be sure to thank a glacier. Without its work sanding down the mountain to create geological "sawdust," we wouldn't see such brilliant colors.

Further Reading

Causes of Color. "What Color is Water?" Accessed March 25, 2019. http://www.webexhibits.org/causesofcolor/5.html.

Gadd, Benn. "Lakes: How Do They Get Those Colors?" In *Handbook of the Canadian Rockies*, 234–36. Jasper, AB: Corax Press, 1992.

Would You Drink This?

Would you drink a mixture of the following ingredients: acetaldehyde, a close chemical relative of the embalming fluid formaldehyde; ethyl acetate, best known as a varnish solvent; acetone, famous as nail-polish remover; acetic acid, also known as vinegar; and a few of the compounds known as hexenals, which give freshly cut grass its characteristic odor?

It sounds horrible. But, in fact, just about all of us have drunk this mixture. These are some of the ingredients of natural grape juice. This list leaves out the three ingredients present in the greatest amounts, namely, water, sugars, and citric acid. But it's the acetaldehyde, ethyl acetate, acetone, acetic acid, and hexenals, among other substances—in small quantities and in the right proportions—that give natural grape juice its characteristic flavor.

Small quantities and correct proportions are important. Take another chemical, hydrogen cyanide, for example. Hydrogen cyanide is naturally present in small amounts in cherries and contributes to their characteristic aroma. But in large quantities, it is a poison.

Another thing you can see from these examples is that the scientific name of a chemical is not likely to tell you whether that chemical is, so to speak, friend or foe.

Which would you rather smell: Hydroxyphenol-2-butanone or trimethylamine? You'd never know from the names alone that the first chemical contributes to the aroma of ripe raspberries and the second causes the stench of rotten fish!

Further Reading

Atkins, P. W. *Molecules*. Scientific American Library. New York: Times Books, 1987.

Tickling the Funny Bone

Of all the parts of our body, the funny bone may have the least appropriate name. That sensitive point on the elbow is not a bone, and it certainly isn't funny.

Most of us have banged an elbow on a sharp corner and felt that indescribable tingling—like an electric shock—up and down the arm. What we are hitting is not a bone but a bundle of nerves, causing them all to fire at once.

Thousands of nerves carry messages from every part of the arm to the brain. Some report on heat, others on cold, and others on pressure. One bundle of nerves passes through a channel in the elbow that we call the funny bone. If you could intercept some of the messages along the way, the messages would all look the same—a combined electrical and chemical impulse. Your brain recognizes heat, cold, and pain only because it knows which nerves sent the signal.

The nervous system is generally reliable, but sometimes it can fool us. Amputees who have lost a leg, for example, may feel pain in the missing foot. The foot is gone, but if the nerves that connected the foot to the brain fire, the brain interprets the signal as pain in the foot.

When you bang the nerves passing through your funny bone against the corner of a table, the shock causes all the nerves to send their messages at the same time. So the message the brain gets is a confused combination of cold, pain, heat, and everything else, which it interprets as coming from all over the arm.

A jolt of electricity can also cause nerves to fire randomly, which is why a bump on the funny bone feels like an electric shock. So banging the funny bone is not very funny, but the reason it feels the way it does is that your brain doesn't know what to think.

The Shape of Snow

Snowflakes are ice crystals, and ice crystals can form as hexagonal plates, needles, or hexagonal columns, as well as the familiar star-like shapes.

All ice-crystal shapes are based in one way or another on the hexagon, the six-sided geometric shape we also see in the cells of a honeycomb. The hexagon is basic to ice crystals because water molecules, when they link up to form ice, take positions corresponding to the corners of a hexagon.

Apparently, just about every snow crystal begins when a tiny amount of water freezes on a piece of dust high in the atmosphere. As this microscopic particle drifts through a cloud, it picks up more and more water molecules from the surrounding humid air. Each new molecule hooks into the existing hexagonal pattern of the crystal. Eventually this process makes an ice crystal big enough to see, with some kind of six-sided symmetry.

Meteorologists have found that the final shape of an ice crystal is dependent on the temperature at which it forms. Below about minus ten degrees Fahrenheit, ice crystals form as hollow hexagonal columns—something like the shape of a pencil. Up to about three degrees Fahrenheit, ice crystals form hexagonal plates; between three and ten degrees, the branching star-shaped types, the so-called dendrites, appear. In warmer air, ice crystals come out as plates, needles, or solid hexagonal columns, depending on the exact temperature.

If an ice crystal drifts through several different temperature regions within a cloud as it forms, it may come out as a hybrid. One of the most beautiful hybrid types is a hexagonal column with a flat plate at each end. These have been named tzuzumi crystals, after the Japanese drums they resemble.

Further Reading

Ahrens, Donald. *Meteorology Today.* 3rd ed. St. Paul, MN: West, 1987.

Knight, Charles, and Nancy Knight. "Snow Crystals." *Scientific American* 228, no. 1 (January 1973): 100–107.

Mason, B. J. "The Growth of Snow Crystals." *Scientific American* 204, no. 1 (January 1961): 120–33.

Remembrance of Things Past for Babies

What do you think is your earliest memory? Maybe when you were three years old looking out the window at a huge hailstorm? That could well be because most people say their earliest memory is from around that age, which led researchers to believe that children start forming long-term memories at around three and a half years old. The thing is, in these studies researchers only asked adults about their earliest memory. Some researchers got to thinking, what would happen if we asked children?

In one study, groups of five-year-olds and eight-to-nine-year-olds were able to recollect memories from their first year of life—some even of their first few months. This implies that the problem isn't that we can't form memories earlier than three and a half, it's just that our brains aren't very good at remembering yet.

In another study, researchers showed three- and four-year-olds a locked treasure chest. Fifteen minutes later, the kids were given a choice among three objects, one of which was a key to the treasure chest. After fifteen minutes, both the three- and four-year-olds were more likely to pick the key. After twenty-four hours, however, only the four-year-olds were more likely to pick the key. The three-year-olds had forgotten all about the chest. The conclusion, then, was that young children's brains can encode information, they just can't recall it later.

Further Reading

Sanders, Laura. "Babies May Be Good at Remembering, and Forgetting." *ScienceNews*, August 28, 2014. https://www.sciencenews.org/blog/growth-curve/babies-may-be-good-remembering-and-forgetting.

Ravens

Avian Einsteins

Ravens are smart and could be called Einsteins of the bird world. They are definitely tops when it comes to relative brain size. And now scientists believe they use objects to get attention.

Ravens are a social species. They form monogamous pairs and are more solitary than crows, but since they don't form breeding pairs until they are three years old, juveniles form flocks or gangs. These social groups help them forage for food. They roost in groups at night and even spend time playing together.

Ravens are known for their large repertoire of calls, feather erections, and body positions to convey anger, fright, affection, curiosity, hunger, and playfulness. But scientists are now finding that they have an additional way to communicate. They apparently use objects to grab the attention of others.

Humans, and apes raised in captivity, use referential gestures to get one another's attention. Pointing is a good example of this type of gesture. It's a good way to initiate a conversation, and scientists believe referential gestures are one of the foundations of developing language.

Ravens don't have fingers with which to make referential gestures, but researchers have documented seven pairs that hold stones, twigs, and moss in their beaks to attract their mate's attention. They only use this behavior when another bird is looking their way, and they only gesture to the opposite sex.

Because pairs defend territory and raise young together, it's an advantage to have good communication skills. More research is needed to determine if this is truly referential communication between pairs or if it's simply a mating ritual. Either way, raven brainpower can't be denied.

Further Reading

Pika, Simone, and Thomas Bugnyar. "The Use of Referential Gestures in Ravens (*Corvus corax*) in the Wild." *Nature Communications* 2, no. 1 (November 2011). doi:10.1038/ncomms1567.

Ant Antennae

Two-Way Communication

You walk into the kitchen one day to find the counter crawling with ants. You wonder what in the world caused them all to be there. Well, it's probably because you left food out. Once one ant finds food, it leaves a pheromone scent trail so others can find the way with their antennae. The antennae are good detectors, so it doesn't take long for the other ants to pick up the pheromone trail.

A really interesting thing is that not only can the antennae pick up information, they can also give information. About 125 years ago, it was first discovered that antennae were used to receive chemical information through touch. For a long time, it seems everyone just assumed that was their only purpose.

However, University of Melbourne scientists decided to study this further, and the results of their research were published in 2016. Ants have a waxy layer covering their body that is made of cuticular hydrocarbon molecules. These unique hydrocarbons protect the ants from dehydration and identify which colony each ant is from. Identification is important so foreign ants or other insects don't invade the colony. Since the antennae are also covered with hydrocarbons, scientists wondered what would happen if they removed the covering just from the antennae.

And sure enough, ants couldn't tell each other's colony identity because ants use only their antenna hydrocarbons to tell other ants which colony an individual is from.

So the question for you is, How do you communicate to the ants that they are in your colony and you want them out of your kitchen? Well, as they say, more research is needed.

Further Reading

University of Melbourne. "Ant Antennae Are a Two-Way Communication System." *ScienceDaily*, March 2016. https://www.sciencedaily.com/releases/2016/03/160330103328.htm.

Wang, Qike, Jason Q. D. Goodger, Ian E. Woodrow, and Mark A. Elgar. "Location-Specific Cuticular Hydrocarbon Signals in a Social Insect." *Proceedings of the Royal Society of London B* (March 2016). doi:10.1098/rspb.2016.0310.

The Echo of a Train

If you stand near a railroad track while a train goes by with its horn blowing, you will hear the pitch of the horn drop as the train passes.

Imagine this sound as waves of slightly compressed air emitted from the horn traveling through the air. The pitch you hear depends on how many sound waves reach your eardrum every second. More sound waves per second means a higher pitch.

As the train approaches, the horn emits each new sound wave when it's a little closer to you than it was for the one before. So the sound waves are crowded together when they get to your ear, and you hear a higher pitch than if the train were standing still.

As the train goes away, the horn emits each new sound wave a little farther from you than the one before; the sound waves are relatively spread out when they get to your ear, and you hear a lower pitch.

So as the train passes, between the time of approaching and the time of going away, the horn's pitch drops: a familiar case of the so-called Doppler effect.

Try listening for this after the train passes: Listen to the horn's echo from surfaces farther down the track. The echo has a higher pitch than the sound directly from the horn. The surface reflecting the sound to make the echo is receiving crowded-together sound waves because the train is approaching that distant surface. That reflecting surface isn't moving, so it doesn't change the pitch before returning the sound to you.

So when a train is going away from you, listen for the horn honking at one pitch, followed by an echo of that honk at a higher pitch.

Old-Fashioned Ice Cream Makers

Have you ever seen an old-fashioned ice cream maker? Nestled inside a wooden bucket, there's a metal canister with a hand crank on top. Cranking the handle scrapes the sweet, creamy mixture from the sides of the metal canister, which sits in a bath of ice water and salt. But what does salt have to do with freezing? Why not just use plain ice water?

A bath of salty ice water freezes ice cream faster than plain ice water. And speedy freezing does more than cut down on cranking time on a hot summer day. A fast freezing process also yields small crystals, ensuring a silky texture. Slower freezing could make big crystals with an unpleasant crunch, kind of like ice cream that's gritty with freezer burn. Here's how salty ice water yields smooth ice cream faster than plain ice water.

The coldest plain ice water can get is thirty-two degrees Fahrenheit, where there's a balance between the rates of melting and freezing. Add salt to the mix, though, and that energy balance tips, allowing the temperature to drop below thirty-two without freezing. Let's say you toss a handful of salt into ice water and the temperature drops to twenty degrees. At this lower temperature, salty ice water can absorb more heat from the mixture inside the canister, and freeze it faster, than plain ice water. And that means you can quit cranking sooner and kick back on the porch to enjoy your silky-textured, old-fashioned ice cream.

Further Reading

Gardiner, Anne, and Sue Wilson. *The Inquisitive Cook*. New York: Henry Holt, 1998.

Wikipedia. "Ice Cream Maker." Last edited June 21, 2019. https://en.wikipedia.org/wiki/Ice_cream_maker.

Forry, Wrong Number

When speaking on the telephone we often run into trouble with particular words that, in face-to-face conversation, aren't troublesome at all. Words such as *feeling* and *ceiling* seem especially hard to get across the line—or to understand when listening to someone else. Is there a reason for this?

In fact, there are several reasons, the major one being that telephones completely remove any visual cues from the process of communication. Not only do we read emotional content off the posture of a speaker's body and face, but language comprehension is aided by a subtle form of lipreading. This has nothing to do with speech or hearing impairments: everyone lip-reads for cues during face-to-face conversations, such as the shape of the lips in producing the *f* sound and not the *s* sound.

A second reason *f* and *s* are so troublesome in phone conversation is that they occur at slightly higher frequencies than telephones are built to transmit. That means the phone simply cuts off the high end of the sound you are trying to produce, making it that much harder to distinguish.

In a sense, telephones replicate the hearing loss that comes naturally with age. Many older people also experience an increasing inability to distinguish the high end of frequencies, making it harder and harder as the years go by to understand other folks' words. So there may be a beneficial side to all this: a young person on the phone can get at least some sense of the frustration an older person may experience every day.

Further Reading

Clason, Debbie. "Understanding high-frequency hearing loss." *Healthy Hearing,* June 22, 2017. https://www.healthy hearing.com/report/52448-Understanding-high-frequency -hearing-loss.

What Obesity and a Lack of
Fatty Tissue Have in Common

Say you're a doctor, and you're seeing a patient that has type 2 diabetes, high blood pressure, and high cholesterol. What do you recommend? Well, since those conditions usually are associated with obesity, you'd probably advise the patient to lose weight, right? The problem is that this patient is skinny: 120 pounds, five foot seven.

For a long time this stumped a lot of doctors. They'd see or hear about those who had all these conditions associated with excess fat but didn't have an ounce of extra fat on their bodies. These patients would also report always feeling hungry and being able to eat seemingly endlessly. It turned out they had a rare genetic disorder called lipodystrophy, which causes an abnormal lack of fatty tissue. The conundrum was this: What was the connection between lipodystrophy and obesity? How could they lead to the same conditions?

Scientists discovered that the connection is the lack of fatty tissue. For obese patients, the lack occurs because all their fatty tissue is already storing fat. For a while, a person just keeps gaining weight, but when their fatty tissue reaches its limit, the body tries to store new fat from excess food in other organs. So the fat goes to places such as the liver, heart, and pancreas. This essentially poisons the body, resulting in the conditions mentioned earlier. Patients with lipodystrophy simply don't have enough fatty tissue to begin with. So any excess food they eat turns into fat that rushes straight to places such as the heart and liver.

Further Reading

Kolata, Gina. "Skinny and 119 Pounds, But with the Health Hallmarks of Obesity." *New York Times*, July 22, 2016. https://www.nytimes.com/2016/07/26/health/skinny-fat.html.

MedicineNet. "Lipodystrophy Definition and Facts." Accessed July 3, 2019. https://www.medicinenet.com/acquired_generalized_and_inherited_lipodystrophy/article.htm#lipodystrophy_definition_and_facts.

WebMD. "Acquired Lipodystrophy." Accessed July 3, 2019. https://www.webmd.com/diabetes/acquired-lipodystrophy#1.

Look through Your Comb at the Mirror

Hold a pocket comb, with teeth vertical, between your eyes and a bathroom mirror. Look through the teeth of the real comb at the teeth of the reflected comb. Slowly move the comb toward the mirror, always keeping both the comb and its reflection in your line of sight. As the comb gets within a few inches of the mirror, you will see what appears to be a shimmering, magnified view of the comb's teeth, with the magnifying power steadily increasing as the comb approaches the mirror.

The shimmering image is a pattern of light and dark created by the overlap of the teeth of the real comb and the reflected comb. In some places, the teeth of the reflected comb fill in gaps between the teeth of the real comb, and you will see solid black. In other places, gaps between teeth on the real comb line up with gaps in the reflected comb. Those areas appear relatively light.

The eerie pattern of dark and light areas is an example of a so-called moiré pattern. Moiré patterns arise whenever two repetitive grid-like designs overlap. In this case, the repetitive design is the row of evenly spaced teeth on the comb. And moiré patterns often resemble a magnified view of the overlapping designs. For example, when you look at overlapping folds of sheer drapery fabric, you see a moiré pattern of crisscrossed dark lines that look like a magnified view of the fabric.

Returning to our comb example, notice that the magnified teeth in the moiré image are even tapered, just like the real ones, and if you point the teeth slightly toward or away from you, the moiré image appears to do the same thing. You have to see it to believe it.

Further Reading

Stecher, Milton. "The Moiré Phenomenon." *American Journal of Physics* 34, no. 4 (April 1964): 247–57.

Blow Out Candles with an Oatmeal Canister

To do this trick, you need an empty cylindrical cardboard oatmeal canister with its lid. Cut a round hole the size of a penny in the center of one end of the canister.

Now aim the canister at a lighted candle, with the hole facing the candle. Tap sharply on the other end. If you have aimed the oatmeal canister correctly, the candle will be suddenly blown out a moment after you strike the canister. With some practice you can blow candles out from up to six feet away.

When you strike the canister, a so-called vortex ring comes out of the hole. This vortex ring is a region shaped like a rubber O-ring constantly turning itself inside out. That turning-inside-out motion of the air enables the vortex ring to retain its shape as it travels toward the candle. That same motion, combined with the forward motion of the vortex ring, blows the flame out when the ring arrives at the candle.

You can make that vortex ring visible as a smoke ring. Fill the oatmeal canister with smoke and then tap very gently on the end of the canister. A smoke ring will emerge from the hole, travel relatively fast for a foot or two, then slow down and spread out. If you tap harder, the smoke ring will travel farther but will be harder to see.

Further Reading

Beeler, Nelson F., and Franklyn M. Branley. *Experiments in Science*. Revised enlarged edition. New York: Crowell, 1955.

Feynman, Richard P. *The Feynman Lectures on Physics*. Reading, MA: Addison-Wesley, 1963.

Wrong Name!

If you're a parent with more than one child, you probably sometimes mix up your kids' names, calling your daughter by your son's name, for example. Or even if you're not a parent, you might remember your parents mixing up your name with that of your brother or sister.

So why do parents do that?

First, to be clear, a parent getting a kid's name wrong has nothing to do with how much the parent loves the child. Mixing up names is a common phenomenon that's caused by how our brains store and recall familiar names. Researchers at Duke looked into this issue by surveying more than 1,700 people ranging from college students to grandparents. And they found some common threads among subjects who had either been called by the wrong name or had called others by the wrong name. They found that people tend to confuse names within relationship groups such as children, grandchildren, friends, and siblings. But the mix-ups only occurred within each group; they didn't cross between groups. So, for example, while it's pretty common for a parent to confuse their kids' names, parents almost never confuse their children's names with the names of their friends.

Why does this happen? Because, the researchers say, related concepts prime each other. Mention "think of a pencil," and you're likely to also think of a pen. The same principle applies to familiar names. In case you're wondering, age-related memory decline doesn't seem to be a factor. College students were just as likely to confuse names as grandparents.

Further Reading

Brusie, Chaunie. "There's Actually a Scientific Reasons We Mix Up Our Kids' Names." Babble. Accessed July 2, 2019. https://www.babble.com /parenting/scientific-reason-mix-kids-names/.

Carroll, Linda. "Why We Mix Up Names of Our Friends and Family." *Today*. January 25, 2017. Accessed July 2, 2019. https://www.today.com /health/mixing-names-family-friends-common-our -brains-t107364.

Steteffler, S. A., C. Fox, C. M. Ogle, and D. C. Rubin. "All My Children: The Roles of Semantic Category and Phonetic Similarity in the Misnaming of Familiar Individuals." *Memory and Cognition* 44, no. 7 (October 2016): 989–99.

When Pop Bottles Don't Blow Up . . . And When They Do

Carbonated beverages in airtight bottles contain dissolved carbon dioxide gas ready to escape when the cap seal is broken. Let a factory-sealed bottle of carbonated beverage sit upright, quietly, all day, then carefully remove the cap. You will hear a puff of escaping carbon dioxide, but the liquid stays put. On the other hand, everybody knows that if you shake the bottle for five seconds and then open it, you get a near explosion with significant and lasting effects on nearby furniture and carpeting.

You might conclude that shaking the bottle increases the pressure inside. But it doesn't. Shaking does create bubbles. The beverage swirls and splashes and falls back on top of itself inside the shaking, sealed bottle. Here and there, carbon dioxide gas gets trapped below the liquid surface, making bubbles in the liquid that were not there before. The pressure is the same, but there are now more bubbles. When you open the bottle, those bubbles suddenly expand because the gas pressure inside them is higher than atmospheric pressure. The rapidly expanding bubbles push liquid out through the bottle's neck.

If you don't shake the bottle, the liquid has few bubbles or none at all. Most of the carbon dioxide gas that's not dissolved invisibly in the liquid is above the liquid, and from there it can escape harmlessly when you loosen the cap.

However, a large and sudden temperature change can indeed alter the pressure inside a pop bottle and cause it to explode. Heating not only causes expansion of the carbon dioxide already present, but also makes more carbon dioxide come out of solution in the pop and rise to the surface as gas, raising the pressure even more.

So as long as the temperature doesn't change, merely shaking a factory-sealed pop bottle will not increase the pressure inside—it just makes more bubbles. When you remove the cap, those expanding bubbles push liquid out of the bottle with a force whose ultimate application depends on your wisdom and judgment.

Further Reading

Deamer, D. W., and B. K. Selinger. "Will That Pop Bottle Really Go Pop? An Equilibrium Question." *Journal of Chemical Education* 65, no. 6 (June 1988). https://doi.org/10.1021/ed065p518.

Common Birthdays
Classic of Probability

Consider a class of thirty children. What is the probability that at least two of them have the same birthday? The surprising answer is that the probability is better than 70 percent that at least two children in a class of thirty have the same birthday.

The secret to understanding this amazing 70 percent figure is to think about the likelihood of all the children's birthdays being different. Imagine asking the children, one at a time, to announce their birthdays. The first child can have any one of 365 different birthdays, of course. The second child can have any one of 364 different birthdays that will not match the first child's birthday. In other words, the chance that the first two children's birthdays will not match is 364 out of 365.

Now the question becomes, what is the chance of getting twenty-nine nonmatches in a row? The third child can have any one of 363 different birthdays that won't match the first two. So the third child's chance of not matching is 363 out of 365. The fourth child's chance of not matching is 362 out of 365, and so on. With each new child, the chance of not matching birthdays with at least one of the previous children gets smaller and smaller. To find the probability of getting nonmatches in a row, you have to multiply all those chances together. A calculator makes it easy. And it turns out that the chance of getting nonmatching birthdays in a row is less than 30 percent. That's why the probability is better than 70 percent that at least two children in a class of thirty will indeed have the same birthday.

Note: The probability of a common birthday in a group of thirty is about 0.7304.

Further Reading

Gamow, George. "The Law of Disorder." In *One, Two, Three . . . Infinity*, 192–230. New York: Viking Press, 1947.

Peters, William Stanley. *Counting for Something: Statistical Principles and Personalities*. New York: Springer-Verlag, 1987.

Take Bets on a Leaky Milk Carton

(Note: Be sure to do this experiment where it won't hurt if things get wet.)

Get an empty half-gallon milk carton. Take a sharp pencil and poke three small round holes in the side of the carton, one above the other: one hole about a quarter of an inch up from the bottom of the carton, a second hole about two inches higher, and a third hole about two inches higher still. Put this milk carton on a flat surface, fill it with water, and watch the water squirt out the three holes. Take bets on this: Which of the three streams will hit the surface farthest from the carton?

You might guess that water will squirt fastest from the bottom hole—and rightly so, because the pressure is greater the deeper you go into a container full of water. Dive to the bottom of a swimming pool and feel the increasing pressure on your ears. So, because the water pressure is greatest behind the bottom hole, the water will squirt fastest from that hole. Therefore, you might guess that the bottom stream will hit the surface farther from the punctured milk carton than either of the other two streams. Try it. You may be surprised to see the bottom stream land closest to the milk carton. Either the middle or the top stream will hit farthest away.

What's wrong? We considered horizontal velocity, but we failed to consider the time needed for the water to fall from the height of each hole. Yes, the water coming out of the bottom hole is moving faster, horizontally, than water from the upper holes. But the bottom stream has less time to fall before it hits the flat surface. During that time the water can travel only a short horizontal distance.

But there is a way to make the demonstration come out as you might think it should. Try this: Move that same leaky milk carton to the edge of the table so the three water streams hit the floor. Now the bottom stream wins; it hits the floor farther from the table than either of the other two.

The distances from the three holes to the surface of impact are now much more nearly the same. The difference in time of fall is small between the top hole and the bottom hole, but the difference in horizontal velocity is still large. Now the difference in horizontal velocity is the crucial factor, and the demonstration comes out as you might expect—the bottom stream wins.

Further Reading

Grimvall, Goran. "Questionable Physics Tricks for Children." *Physics Teacher* 25, no. 6 (September 1987). https://doi.org /10.1119/1.2342286.

Paldy, Lester G. "The Water Can Paradox." *Physics Teacher* 1, no. 3 (September 1963). https://doi.org/10.1119/1.2350617.

Smells and Memories

When you step outside and sense the transition from autumn to winter or notice signals of a fast-approaching spring, you likely experience a feeling of being transported back in time and place. Perhaps childhood memories of shoveling snow or kindling a campfire resurface with vivid intensity. Sensory stimuli have the power to involuntarily trigger such memories. Olfactory stimuli—cues we detect through smell—are special because they activate parts of your brain that spark memories that are often acutely emotional.

A good way to comprehend this phenomenon, and also learn more about your own memories, is to understand how anatomical pathways process olfactory stimuli. After a smell enters the nose, a neural structure called the olfactory bulb relays this odor information to the brain. The information travels through two cerebral regions: the amygdala and the hippocampus, which deal with emotion and memory respectively. Auditory, visual, and tactile information do not take this route to the brain, however, which means information from these senses is less likely to induce feelings of "being taken back" or revisiting a past emotional state.

As we learn more about what happens when smell stimulates memory, it becomes clearer that such memories are likely to be emotionally charged. On one hand, odor-evoked memories increase temporal lobe activity, which has been linked with positive memory processing, thus creating more pleasant experiences than memories brought on by visual cues.

On the other hand, people living with PTSD may experience certain smells as potent triggers that can, without warning, shuttle them back to a moment of trauma. What might your recollection of certain odors tell you about your past?

Further Reading

Arshamian, Artin, Emilia Iannilli, Johannes C. Gerber, Johan Willander, Jonas Persson, Han-Seok Seo, Thomas Hummel, and Maria Larsson. "The Functional Neuroanatomy of Odor Evoked Autobiographical Memories Cued by Odors and Words." *Neuropsychologia* 51, no. 1 (Jan 2013): https://doi.org/10.1016/j.neuropsychologia.2012.10.023.

Bergland, Christopher. "The Neuroscience of Smell Memories Linked to Place and Time." The Athlete's Way. *Psychology Today*, July 31, 2018. https://www.psychologytoday.com/us/blog/the-athletes-way/201807/the-neuroscience-smell-memories-linked-place-and-time.

Gaines Lewis, Jordan. "Smells Ring Bells: How Smell Triggers Memories and Emotions." Brain Babble. *Psychology Today*, January 12, 2015. https://www.psychologytoday.com/us/blog/brain-babble/201501/smells-ring-bells-how-smell-triggers-memories-and-emotions.

White, Jess. "A Hint of Memory." *Psychology Today*, May 7, 2016. https://www.psychologytoday.com/us/articles/201605/hint-memory.

Wikipedia. "Olfactory Memory." Last edited May 2, 2019. https://en.wikipedia.org/wiki/Olfactory_memory.

Big Shadows

Try this after a candlelight dinner: Blow out all the candles but one, then turn off all the other lights in the room. Look at the walls. The candle will throw frighteningly huge shadows of you and your friends onto the walls. The bigger the room, the bigger the shadows.

Being careful not to get burned, hold your hand about two feet from the candle and notice the size of the shadow your hand casts on the wall. Now move your hand to a position one foot from the candle; notice how much larger the shadow becomes. There's a mathematical proportion in this situation. If your hand is five times closer to the candle than to the wall, the shadow of your hand will be five times bigger than your hand itself. If your hand is ten times closer to the candle than to the wall, the shadow will be ten times as big as your hand. And so on.

Why doesn't an ordinary living-room electric lamp cast huge, dark shadows? The answer is that electric lamps and most electric light bulbs are designed to cast soft shadows, unlike a bare candle flame. The candle on the dining-room table is a small light source; the flame is usually less than an inch high. A frosted light bulb, on the other hand, is several inches high and gives off light from all over its surface—it's a bigger light source than a candle. A lampshade makes the effective size of the light source even bigger. The larger the light source, the more diffuse the shadow. If you put your hand a foot away from a candle, you block a large amount of the light that would otherwise reach the walls. If you put your hand a foot away from an electric lamp with a frosted bulb and a big shade you block a much smaller proportion of the light, so the shadow on the wall is much less noticeable.

Further Reading

Lynde, C. J. *Science Experiences with Ten-Cent Store Equipment.* 2nd ed. Princeton, NJ: Van Nostrand, 1950.

Half Heads, Half Tails

Flip an honest coin, and you expect a head just as much as you expect a tail. Flip a coin ten times, and you expect to get about half heads and half tails. But here's an interesting question: Should you expect to get exactly five heads and five tails?

The answer is no. It is very likely that ten flips will give approximately half heads and half tails. But it is much less likely that ten flips will give exactly half heads and half tails.

Here are some more precise numbers, calculated from the laws of probability. If you flip a coin ten times, there is less than a 25 percent chance of getting exactly five heads and five tails. However, there is almost a 66 percent chance that you'll be close to five heads and five tails; in other words, there's almost a 66 percent chance that you'll get either four heads and six tails, five heads and five tails, or six heads and four tails.

The more times you flip the coin, the more likely it is that you will get approximately half heads and half tails, but the more times you flip the coin, the less likely it is that you will get exactly half heads and half tails. If you flip a coin twenty times, your chance of getting exactly ten heads is only about 18 percent, but your chance of getting something between eight and twelve heads is about 74 percent.

If you flip a coin a million times, your chance of getting exactly 500,000 heads is infinitesimally small, but your chance of getting between 499,000 and 501,000 heads is extremely large.

Note: The numbers come from the binomial distribution.

Further Reading

Huff, Darrell. *How to Lie with Statistics*. New York: Norton, 1954.

Spiders Don't Get Caught in Their Own Webs

Spiders have an oily secretion on their feet that keeps them from sticking to their webs. But there's more to the story than that. Not all the threads made by a spider are sticky. A spider can make more than one kind of silk. Spiders use their silk not only to make the sticky parts of their webs, but to line their burrows, to wrap their eggs, and as parachutes enabling them to travel on the wind.

Even in a web whose function is to trap insects, not all the threads are sticky. Take, for example, the so-called orb web—the kind that probably comes to mind first when we think of a spider web. An orb web has one set of threads emanating from the center and another thread applied over those radial threads in a spiral pattern something like the groove on a phonograph record. Generally, only the spiral thread is sticky, not the radial threads.

But even the sticky threads in a web are not sticky over their entire length. A spider makes a sticky thread by applying glue to the silk as it is spun. The spider applies glue continuously to the new thread as it emerges from the end of the spider's abdomen.

This glue doesn't form a continuous coat on the silk thread, however. Surface tension comes into play.

Surface tension is the same attraction between molecules that makes water gather into beads on a well-waxed car. Surface tension in the spider's glue gathers that glue into beads, arranged along the thread like beads on a string. Those beads of glue on the spiral threads of an orb web may be just barely visible through a good magnifying glass.

Further Reading

Bristowe, W. S. *The World of Spiders.* London, UK: Collins, 1971.

Grzimek, Bernhard. *Grzimek's Animal Life Encyclopedia.* New York: Van Nostrand Reinhold, 1984.

Zim, Herbert S. *Spiders and Their Kin.* New York: Golden Press, 1990.

Bilingual Brain

Have you ever wondered what it would be like to speak more than one language? Would it be hard to talk without mixing up the languages?

Here's what scientists have to say: Using two languages provides the brain with plenty of practice exercising a certain kind of cognitive control. When a person is bilingual, for example, he or she has two different languages constantly active in the mind. In order to speak one language without intrusions from the other, the speaker's brain needs to suppress the language that isn't being used.

So how exactly does the brain manage to hold back one language in favor of another? Researchers have argued that the same mental processes and skills that allow us to do things such as focus our attention, remember directions, and control impulses work in the bilingual brain to suppress the language that isn't being used. In fact, studies seem to show that thanks to the practice the brain receives while managing two languages, bilingual folks may gain some mental advantages even beyond the realm of language control. Researchers found that a lifetime spent speaking two languages appears to slow the rate of decline for some mental processes as people age, and other studies suggest advantages that have to do with creativity and problem-solving.

So, based on this research, it seems that using more than one language may be good for the brain.

Further Reading

Bialystok, E., F. M. Craik, R. Klein, and M. Viswanathan. "Bilingualism, Aging, and Cognitive Control: Evidence from the Simon Task." *Psychology and Aging* 19, no. 2 (June 2004). https://www.ncbi.nlm.nih.gov/pubmed/15222822.

Center on the Developing Child at Harvard University. "Executive Function and Self-Regulation." Accessed July 2, 2019. https://developingchild.harvard.edu/science/key-concepts/executive-function/.

The Shape of the Earth

Ancient civilizations described the earth as a flat plate, a dome, or even a huge drum. Today, with the help of photographs from space, we know the earth is a sphere. Yet even by the fourth century BC, Greek astronomers with no more evidence than you could collect in your own backyard believed the earth was spherical.

Aristotle, for example, based his argument for a spherical world on the shape of the earth's shadow during a lunar eclipse. A lunar eclipse occurs when the sun and the moon line up on opposite sides of the earth. When that happens, the shadow of the earth moves across and eventually covers the moon. Although the moon is too small to show us the entire shadow of the earth, the edge of the shadow we can see is always curved, regardless of where the moon and sun are in relation to the surface of the earth. The only shape that could cast a round shadow from every angle is a sphere. A round, flat plate would cast a round shadow only if the light were coming from above or below; from any other angle, the shadow would be an oval or a straight line.

Anyone driving west across the country will notice that the peaks of the Rocky Mountains appear before the foothills. If we say this is because the foothills are still below the horizon, we are implying—quite correctly—that the curvature of the earth gets in the way. At closer range, there is no horizon between us and what we are looking at, so the earth appears flat. But if the earth were really flat, the only limit to how far we could see would be the power of our vision, and on a clear day the full height of the Rockies would be visible from Kansas.

Further Reading

Cohen, Morris R., and I. E. Drabkin. *A Source Book in Greek Science*. Cambridge, MA: Harvard University Press, 1958.

Kuhn, Thomas S. *The Copernican Revolution*. Cambridge, MA: Harvard University Press, 1957.

It's Not What You Hear—
It's When You Hear It

You're sitting near the back of a big auditorium, and someone on the stage is talking. If you can hear the talker clearly, there's probably a sound-reinforcement system at work—a system involving microphones, amplifiers, and loudspeakers.

In some auditoriums, the talker's voice is fed not into one loudspeaker at the stage but into many loudspeakers mounted in the ceiling or along the side walls of the room. This approach puts a loudspeaker near everyone in the audience. But this approach creates a problem. Electronic signals can travel almost instantly to the back of a big room, while the live sound takes a noticeable fraction of a second to travel to the back through the air. So the best sound-reinforcement systems delay the electronic signal by a fraction of a second before sending it to loudspeakers at the back.

How long should that delay be? It might seem best for the amplified sound to reach you through the loudspeakers at the exact moment live sound reaches you through the air. But audio engineers and psychologists have found that, for the most natural sound, the delay has to be just a little longer than that— maybe a fiftieth of a second longer.

This is because of a so-called precedence effect in human hearing. If the live sound reaches your ear about a fiftieth of a second before the amplified sound, all the sound will appear to come from the stage, not from speakers in the ceiling. This is true even if the loudspeaker sound is somewhat louder than the live sound!

Sound appears to come from whichever source produces it first. If the delays in a sound-reinforcement system are set just right, you may never be aware that amplification is being used at all.

Further Reading

McGraw-Hill Encyclopedia of Science and Technology, s.v. "Sound-Reinforcement System." New York: McGraw-Hill, 1987.

Olson, Harry F. *Music, Physics and Engineering*. New York: Dover, 1967.

Weightless Water

Punch a hole somewhere near the bottom of an empty tin can. Fill the can with water. Of course, a stream of water squirts out of the hole. Now refill the container and drop it from a height of five or six feet. Notice that while the container is falling, water does not squirt out of the hole. The water is weightless with respect to the can as long as it is falling.

You can see how this happens by remembering that the force of gravity makes all falling objects—water, cans, and everything else—accelerate toward the ground at the same rate. (We're neglecting air resistance here because it doesn't affect the main point.) When you're holding the punctured can, it cannot accelerate toward the ground because it's not free to move. But the water presses against the sides of the container, escapes through the hole, and accelerates toward the ground.

When you drop the punctured can, gravity makes the can and the water accelerate toward the ground at the same rate. Gravity pulls on the water, but it also pulls on the can. So the weight of the water does not press on the sides and bottom of the can. The water is weightless and no longer squirts through the hole.

The water-filled can does not have to fall straight down for this demonstration to work. You might throw it sideways, for instance. However you throw it, once the can leaves your hand, no water squirts through the hole.

You might want to ponder this: If you could throw the can sideways at seventeen thousand miles an hour, it would be going as fast as an orbiting space shuttle. Again, the water inside would be weightless—and for the same reason the shuttle astronauts are weightless in orbit.

Further Reading

Kutliroff, David R. *101 Classroom Demonstrations and Experiments for Teaching Physics*. West Nyack, NY: Parker, 1976.

The Force of a Tornado

Tornadoes are known for their powerful winds, but the winds alone cause only part of the damage. In addition, the high winds create an area of extremely low pressure that can actually cause buildings to explode.

To understand how the winds cause the pressure to drop, we can review Bernoulli's principle, the same principle that explains the curveball. Bernoulli's principle states that a fast-moving air stream has lower pressure than slow-moving or still air around it. So when the very high winds of a tornado blow over a house, they cause the pressure to drop to a level far below that of the still air inside the house. The higher pressure in the house can then force the roof off the top of the house. As the roof comes up, the tornado winds pick it up and carry it away.

To see how this works on a much smaller scale, hold a dollar bill on the back of one hand. With the other hand, hold the edge that is away from you so the bill doesn't fall off. Now, blow hard across the top surface of the bill, and it should lift up off your hand. If you blow down onto the paper it won't work, but if you blow parallel to the surface of the bill, you should be able to decrease the air pressure above the paper enough to cause it to rise. Once it rises slightly, your breath will catch the underside of the bill and blow it back across your other hand.

When a window is broken in a tornado, the glass usually lands outside the house. This is because the force is exerted from the area of higher pressure inside the house toward the area of lower pressure outside the house.

Further Reading

Ahrens, C. Donald. *Meteorology Today: An Introduction to Weather, Climate, and the Environment.* St. Paul, MN: West Publishing, 1991.

The QWERTY Effect

Today, most writing happens through an interface with technology. The act of writing is now synonymous with the task of typing as more work-related and personal correspondence takes place at the computer keys. Researchers David Garcia and Markus Strohmaier have traced the rise of an interesting phenomenon connected to this integration of the keyboard as our primary writing tool.

The phenomenon is called the QWERTY effect, which is named after the first six letters at the top left of the keyboard. It describes the finding that words with higher ratios of letters from the right side of the keyboard are linked to more positive emotions than words that use more letters from the left side of the keyboard. The researchers demonstrated the idea with a sample of English speakers, all of whom ranked words with higher proportions of letters from the right keys as more positive.

But why does this happen? Previous research noted that regardless of whether they are left- or right-handed, people on average type faster with the right hand. Also consider that there are fewer *letters* on the right side of the keyboard. These factors make typing with the right hand easier, which over time has conditioned our perceptions of the letters themselves.

Some researchers have remained unconvinced by these findings, but Garcia and Strohmaier continue to produce supporting evidence. For example, they've found that across websites like YouTube, Amazon, and Rotten Tomatoes products and titles with higher ratios of letters from the right side of the keyboard tend to have higher ratings. In the future, it would be interesting to study if the QWERTY effect contributes to the popularity of certain online messages or posts.

Further Reading

Baraniuk, Chris. "The layout of QWERTY Keyboards Shapes Our Feelings About Words." *New Scientist*, April 20, 2016. https://www.newscientist.com/article/2085334-the -layout-of-qwerty-keyboards-shapes-our-feelings-about -words/.

Mosher, Dave. "The QWERTY Effect: How Typing May Shape the Meaning of Words." *Wired*, March 7, 2012. https:// www.wired.com/2012/03/qwerty-effect-language/.

The Spinning Earth and the Weather

You and a friend sit on opposite sides of a big, flat turntable—a merry-go-round. With the turntable not moving, you toss a ball to your friend on the other side. If you toss accurately, the ball goes to the other person, who catches it.

Now the turntable begins to spin counterclockwise. Because of the turning, you begin moving to your right. You toss the ball to your friend on the other side once again. If you aim as you did before, you'll miss; the ball will turn right from your point of view. Your friend on the other side will have to lunge to his or her left to catch the ball. Because you're moving, the velocity of the turntable combines with the velocity of your throwing arm to send the ball off to the right. (If the turntable were spinning clockwise, the ball would turn left.)

This effect was first analyzed in detail in 1835 by the French physicist Gaspard Gustave de Coriolis, who was making a theoretical study of the forces on moving parts of machines. Now we understand the tremendous importance of the Coriolis effect, as it has come to be called, in explaining how things move long distances over the earth—air masses, for instance.

The earth is, in effect, a giant turntable. As seen from the North Pole, the earth spins counterclockwise every twenty-four hours. Because of that spinning, air flowing out from a northern-hemisphere high-pressure area turns right, just like the ball leaving your hand, and that causes clockwise wind circulation.

The Coriolis effect of the earth's rotation is noticeable only with things traveling very long distances—things like winds and ocean currents. Contrary to what we sometimes hear, it's too weak to have a noticeable effect on water going down the drain in a sink.

Further Reading

Ahrens, C. Donald. *Meteorology Today.* 2nd ed. St. Paul, MN: West, 1985.

Marion, J. B. *Classical Dynamics of Particles and Systems.* 2nd ed. Orlando, FL: Academic Press, 1970.

The Floating Cork Trick

Cut a slice about a quarter of an inch thick from the end of a cork and drop it into a glass of water. Watch the cork for a few seconds, and you'll see it drift over to the side of the glass. Challenge everyone at the table to make the cork float exactly in the center of the water. Let people push the cork around, turn it over, drop it in in some special way. No matter what they try, it will always drift to one side . . . for the uninitiated.

The trick is to leave the cork in the glass and add more water. Pour it in slowly from another glass. Keep pouring slowly and carefully until the water surface bulges above the rim of the glass itself. You may be surprised at how much water will fit into that bulge without spilling.

The water bulges because molecules of water attract each other. The molecules at a water surface make a film under tension. That film of surface tension holds the water like a bag and keeps it from spilling. The same thing happens in a water drop.

Bring your eye level with the rim of the glass and look at the profile of the water. You'll notice that it curves gently across the mouth of the glass, with the highest point of the bulge at the center. By this time you'll also notice that the buoyancy of the cork has caused it to be pulled up the sloping water surface to that central high point.

Why didn't the cork float in the center before the glass was full? Pour some water from the glass and look carefully at the surface. Now you won't see the bulge in the center, but you can see that the water climbs up the sides a little all around the edge. This is because the water molecules are attracted to the glass.

When the glass is not full, the water is higher at the edges than in the center, so the buoyancy of the cork still causes it to float to the highest point, but that is now at the edge rather than the center.

Further Reading

Gardner, Martin. *Entertaining Science Experiments with Everyday Objects*. New York: Dover, 1981.

On a Clear Day, How Far Can You See?

How far you can see depends on the condition of the atmosphere and on whether anything is blocking your view. But if we assume the air is perfectly clear and the horizon is unobstructed, how far can you see?

Because the earth is round, the higher you are, the farther you see. In case you're the calculating type, here's the formula: Multiply the square root of your height, in feet, by a factor of one and a quarter. That gives the approximate number of miles to your horizon.

In case you're not the calculating type, here are some results:

- If your eyes are five feet above ground level, your horizon is about two and three-quarters miles away.
- If you're on the tenth floor of a tall building—about a hundred feet up—your horizon is about twelve miles away.
- From fourteen hundred feet up—that's roughly the height of the Empire State Building—you can see forty-six miles if the air is perfectly clear.
- From a jet at thirty thousand feet you can see about 210 miles; from a spacecraft a hundred miles up, about 890 miles.

That formula once again: Multiply the square root of your height in feet by a factor of one and a quarter. That gives you the approximate number of miles to your horizon—on Earth.

Elsewhere, the formula changes. On the moon, for instance, if your eyes are five feet off the ground, you can see only about 1.4 miles. Since the moon is smaller than the earth, the surface curves more sharply, and the horizon is closer.

If you could stand on the sun, you'd be standing on a surface more nearly flat. From five feet above the sun, with nothing blocking your view, you'd see thirty-one miles.

The flatter the surface, the farther you see. If the earth were absolutely flat, on a clear day you really could see forever.

Further Reading

Bakst, Aaron. *Mathematics: Its Magic and Mystery*. 3rd ed. Princeton, NJ: Van Nostrand, 1967.

Benjamin Franklin and the
Swatches on the Snow

In a letter written in 1761, Benjamin Franklin tells how he collected some little squares of broadcloth, tailor's samples of different colors: black, dark blue, light blue, green, purple, red, yellow, and white. He wanted to demonstrate that these colors would absorb different amounts of light from the sun and convert the light to different amounts of heat. On a bright winter day, when the ground was blanketed with freshly fallen snow, Franklin laid the cloth squares on the snow in the sun and left them for a few hours. When he came back, he saw that the black square had sunk deeper into the snow than any of the others. The dark blue cloth had sunk a little less, and the white square not at all. Each of the other squares had melted its way down to an intermediate depth.

Even two hundred years ago, people knew that dark-colored things get warmer in the sun than light-colored things. But Franklin's little experiment demonstrated it scientifically by comparing cloth samples that were all the same except for one thing: color.

The dyes in the samples absorbed different amounts of sunlight. The black cloth got the warmest because it absorbed all the colors of the sunlight and reflected almost none—that's why it looked black. The white cloth stayed the coolest; it reflected all the colors of the sunlight and absorbed very little light. The red cloth absorbed some of the sunlight and warmed to a medium temperature, but it reflected red light to the eye and looked red. And so on for the other colors.

Benjamin Franklin did other experiments, especially with electricity, and even speculated about whether flies drowned in wine could be brought back to life. But that's another story.

Further Reading

Goodman, Nathan G., ed. *The Ingenious Dr. Franklin: Selected Scientific Letters of Benjamin Franklin.* 1931. Reprint, Philadelphia, PA: University of Pennsylvania Press, 1974.

Dog Facial Expressions and Humans

Many mammals make facial expressions. The architecture of the face, with muscles for making facial expressions, is much the same across these mammals. They probably inherited this architecture from a common evolutionary ancestor in the distant past. Scientists used to think that animal facial expressions aren't like ours, that they are inflexible and involuntary and don't reflect sophisticated cognitive processes.

However, in 2017 a team of scientists from the United Kingdom published findings that showed dogs actively use facial expressions to communicate with and influence humans. The experimenters put dogs in situations where a human experimenter was either paying attention to them or turned away and either gave them food or didn't.

It probably comes as no surprise to dog owners that getting food is emotionally exciting for dogs, and you've probably seen the anticipation in their faces. But if their facial expressions just have to do with emotional excitement, it shouldn't matter whether a human is paying attention to them or not. On the other hand, if dogs are actively using facial expressions to influence humans, it should.

And in fact, it did. The researchers found that the dogs showed a much wider range of facial expressions when the human was paying attention than when they were given food without human attention.

But there was more. Researchers found that dogs use a special gesture involving raised eyebrows when interacting with humans. It makes their eyes look bigger, which humans find cute. Dogs that use this expression might be more likely to be adopted from a shelter.

Further Reading

Davis, Nicola. "Dogs Have Pet Facial Expressions to Use on Humans, Study Finds." *Guardian*, October 19, 2017. https://www.theguardian.com/science/2017/oct/19/dogs-have-pet-facial-expressions-to-use-on-humans-study-finds.

IFL Science. "Study Proves That Your Dog Really Is Deliberately Manipulating You." Accessed July 3, 2019. http://www.ifl science.com/plants-and-animals/your-dog-only-uses-puppy-dog-eyes-when-it-knows-you-are-looking/.

Kaminski, Julianne, Jennifer Hynds, Paul Morris, and Bridget M. Waller. "Human Attention Affects Facial Expressions in Domestic Dogs." *Scientific Reports* 7 (2017). doi:10.1038/s41598-017-12781-x.

Potenza, Alessandra. "Dogs Make More Faces When You're Paying Attention." *Verge*, October 19, 2017. https://www.theverge.com/2017/10/19/16497078/dogs-facial-expressions-social-bond-interaction-people.

Starr, Michelle. "Dogs Really Do Pull Faces to Communicate with Us, Says New Study." *Science Alert*, October 20, 2017. https://www.sciencealert.com/dogs-facial-expressions-cognition-human-communication-october-2017.

Why Is the Sky Blue?

Why is the sky blue? It can't be that the atmosphere has a blue color like the blue of a tinted windshield. In that case, going outside in the daytime would be like walking around inside a blue glass bottle, with a blue sun shining blue light everywhere, and blue stars and a blue moon at night.

The blue can't be from dust because the air over gravel parking lots and quarries is whitish, not bluish.

The blue can't be the result of water droplets. Clouds are made of water droplets, and clouds are white. It's not a matter of relative humidity, either; a dry sky over Arizona can be just as blue as a humid sky over Minnesota.

Blue is not the color of outer space. The background of space is black. So the sky is black at night—it's blue only during the daytime, when the sun is shining on the atmosphere.

The sun shines with all the colors of the rainbow—blue, yellow, red, and all the rest—mixed together to make white light. The reds and yellows pass through air easily, but some of the blue portion of sunlight is scattered in every direction by air molecules. When you look to the sky on a clear day, you see blue light scattered from sunbeams by molecules of nitrogen, oxygen, and carbon dioxide in Earth's atmosphere.

The more air, the more scattering. In early morning and late afternoon, the sun's light passes through so much air that most of the blue has been scattered away by the time the light reaches you. The reds and yellows remain, and the sun looks reddish.

Further Reading

Feynman, Richard P. *The Feynman Lectures on Physics*. Reading, MA: Addison-Wesley, 1964.

Minnaert, Marcel. *The Nature of Light and Colour in the Open Air*. New York: Dover, 1954.

Why One Rotten Apple
Can Spoil the Barrel

If you buy green, underripe lemons or bananas, you can make them ripen faster by keeping them in a paper bag. Chinese growers used to ripen fruit by keeping it in a room with burning incense. In the West, farmers used to "cure" fruit with kerosene stoves.

What's behind all these processes is a gas called ethylene. Ethylene comes from burning fuels like kerosene, and it's made naturally by all parts of a plant at one time or another. Ethylene stimulates germination of the seed, flowering, ripening of fruit, dropping of fruit, and dropping of leaves. The ethylene produced by ripe fruit will stimulate another nearby fruit to ripen. So there's a chemical basis for the proverb about one rotten apple spoiling the whole barrel.

Ethylene is used in commercial agriculture to stimulate the ripening of bananas, tomatoes, and citrus fruits and to help give them the colors people expect.

Fruit shippers usually want to stop the ripening process, so they store fruit in rooms that have ethylene chemically removed from the air. Incidentally, farmers at one time thought it was the heat from kerosene stoves that was ripening their fruit. But those who tried modern nonkerosene heaters didn't get the results they wanted. It wasn't the heat that was "curing" the fruit—it was the ethylene.

Further Reading

McGraw-Hill Encyclopedia of Science and Technology, s.v. "Ethylene." New York: McGraw-Hill, 1987.

Salisbury, Frank B., and Cleon W. Ross. *Plant Physiology*. 3rd ed. Belmont, CA: Wadsworth, 1985.

Diamonds

You walk into a fancy party. What is the oldest thing there? No, not your in-laws, or even your grandparents. It will be a diamond someone is wearing. More on the age of diamonds later, but first some more amazing mysteries and facts about diamonds.

It's not completely clear how natural diamonds are formed, and although it is known that they are made from carbon, scientists are not sure how carbon gets to the extreme depths at which diamonds are made.

Diamonds are formed very deep in the earth. The optimum depth for their formation is about 120 miles beneath the surface, in the molten mantle. The temperature and pressure necessary for diamonds to crystallize are mind-boggling: the temperature must be over two thousand degrees Fahrenheit, and the pressure must be at least 690,000 pounds per square inch. To put that pressure in perspective, a 150-pound person exerts only about 3 pounds per square inch. We would never know about diamonds were it not for volcanic activity, for that's what delivers diamonds from deep inside the earth to where we can get to them.

Theoretically, diamonds can remain diamonds only at high temperature and pressure. In theory, at atmospheric pressure and lower temperature, chemical changes are liable to take place that change diamond into graphite, similar to the stuff pencil leads are made of. However, it has been calculated that even if this change were to take place, it would take more than ten billion years.

Now back to the age of diamonds. Scientists think diamonds may have been forming throughout earth's history. Many have

been found that are 3.3 billion years old. And the young ones are
a mere one billion years old.

Bibliography

Kirkley, M. B., J. J. Gurney, and A. A. Levinson. "Age, Origin
and Emplacement of Diamonds: A Review of Scientific
Advances in the Last Decade." *CIM Bulletin*, January 1992.

Legrand, Jacques. *Diamonds: Myth, Magic, and Reality.* New
York: Crown Publishers, 1980.

Wilson, A. N. *Diamonds from Birth to Eternity.* Santa Monica,
CA: Gemological Institute of America, 1982.

Saccadic Suppression

Here's a fun demonstration you can do at home. Look in the mirror, and shift your focus from one eye to the other. Can you see your eyes move? No, you can't; it's as if they don't move at all. Then have someone face you and shift their focus from one of your eyes to the other. You can see their eyes move, so why can't you see your own eyes move? You're not the first person to wonder about that; it could be that people have pondered this as long as there have been mirrors.

So, in the mid-1970s scientists did a study to answer this question. Rapid eye movements are called saccades, and scientists studying them found that our vision is suppressed right before and during each saccade. It's called saccadic suppression. The researchers explained that by suppressing our vision whenever our eyes shift rapidly from one point to the next, our brain creates perceptual stability. Otherwise the world around us would seem to move every time we moved our eyes. That would be disorienting, to say the least.

Further Reading

Bridgeman, Bruce, Derek Hendry, and Lawrence Stark. "Failure to Detect Displacement of the Visual World during Saccadic Eye Movements." *Vision Research* 15, no. 6 (1975). https://doi.org/10.1016/0042-6989(75)90290-4.

Krekelberg, Bart. "Saccadic Suppression." *Current Biology* 20, no. 5. Accessed July 3, 2019. https://www.cell.com/current -biology/pdf/S0960-9822(09)02137-X.pdf.

Mathôt, Sebastiaan. "Can You See While Your Eyes Move?" *COGSCIdotNL: Cognitive Science and More*, August 6, 2015. http://www.cogsci.nl/blog/miscellaneous/242-can-you-see -while-your-eyes-move.

Wikipedia. "Saccadic Suppression of Image Displacement." Last edited July 15, 2018. https://en.wikipedia.org/wiki /Saccadic_suppression_of_image_displacement.

Spoonerisms

At some point, everyone has transposed the first letters in two words and come up with a nonsense phrase. You might mean to say *barn door*, but it comes out *darn boor*. These slips of the tongue are called spoonerisms, and cognitive psychologists study them because of what they say about how our brains construct language.

Early twentieth-century psychologists believed language was produced in our brains one word at a time and each word acted as a stimulus to produce another word. But cognitive psychologists now believe we produce language in clumps rather than one word at a time.

The study of spoonerisms has helped scientists formulate these new theories. Spoonerisms may seem like random mistakes, but in fact they follow a regular set of rules. When two sounds are transposed between two words, they are almost always sounds that belong in the same position. For example, the beginning of one word almost never exchanges with the end of another. The close association your brain makes between two words such as *barn* and *door* indicates your brain chose those words as a unit, rather than one at a time.

We make speech errors like this because as we construct language, our brain builds a frame for what we are going to say before we choose the actual words that will go into that frame. When we get a phrase right, our brains have successfully coordinated this frame with the sound of a word. Spoonerisms happen when this coordination breaks down, often because of the interference of an external or internal stimulus.

Dimples in Golf Balls

You're out on the golf course one pleasant afternoon. Your ball is set on the tee. You lean over the ball. You grip the club just right. Your arms are straight. Your stance is perfect. Whoosh! You swing, and the ball takes off toward the green, sailing in a beautiful arc. Your impeccable technique certainly has a lot to do with the success of that drive, but you got a major assist from the little dimples on the golf ball. Combined with the proper spin, the dimples help keep the ball in the air longer, and here's how.

Your picture-perfect swing puts backspin on the ball. The dimples trap a layer of air next to the ball, and this layer spins with the ball. The air being dragged across the top of the spinning ball moves in the same direction as the air that's rushing past. As the air spinning with the ball comes around the bottom, it is moving in the opposite direction from the air on top, and therefore against the onrushing air. Consequently, it's slower than the air on top. Once again we encounter Bernoulli's principle: a slow-moving air stream has higher pressure than a fast-moving air stream. So the higher pressure on the bottom of the ball is going to hold the ball up longer. The effect of the dimples is so significant that a drive of two hundred yards hit with a dimpled golf ball would be shortened with a nondimpled ball to about one hundred yards.

Incidentally, with topspin, which is the reverse of backspin, the principle is the same, but the effect is just the opposite—and disastrous for the golfer: with topspin the high pressure is on the top of the ball, so after a short flight the ball takes a nosedive into the ground.

Further Reading

Flatow, Ira. *Rainbows, Curve Balls and Other Wonders of the Natural World Explained.* New York: William Morrow, 1988.

Why Do Cats' Eyes Glow at Night?

You're driving along a lonely road at night. Ahead, in the dark, you see a pair of bright, disembodied eyes. You get closer and the eyes slip away into the grass—the eyes of a prowling cat.

The cat's eyes shine because your eyes, your car's headlights, and the cat's eyes are nearly in a straight line. The light from your headlights, or the sound from your engine, attracts the attention of the cat toward your car. The cat focuses its eyes on your headlights. In each of the cat's eyes, the lens brings the light from your headlights to a sharp focus, making a distinct image on the retina like the image on film in a camera.

But the light goes both ways. Some of it is reflected from the cat's retina back out through the lens of the eye. At night, the pupil of the cat's eye is wide open, so a lot of light goes through. The reflected light forms a narrow beam aimed at your car because that's where the cat is looking. You look into the beam and get the weird impression that the eyes are shining by their own light. The impression is even weirder if you can't see the rest of the cat.

You can see the same glow in the eyes of rabbits that look up from their nighttime grazing to watch your car go by. And you can see it in the eyes of people in snapshots taken at evening parties. Anyone looking at the camera when the flash goes off has red eyes in the picture. The red comes from the color of the human retina.

This is easy to demonstrate with a model of an eye: Put a lens over a hole in a box whose depth is about the same as the focal length of the lens—the focal length should be marked on the metal barrel of the lens, for example 50 mm. Stand at a distance and shine a flashlight beam on the box.

But there's more. Cats have a special layer of tissue in their eyes that reflects light somewhat like a metallic surface does. It is located just behind the special cells in the retina that convert light to nerve impulses—the rod and cone cells. The apparent function of this reflecting layer is to make the cat's eye more efficient in dim light.

When light enters the cat's eye, some of it is absorbed by the rods and cones. But some light gets past the rods and cones and goes on to the back of the eye. The reflecting layer bounces this leftover light forward again, so it encounters the rods and cones a second time. Light entering a cat's eye has not one but *two* chances to be detected by the cat.

Some light misses the rods and cones both times and leaves the cat's eye through the lens. That's the light we see as eyeshine. But because of this reflecting layer in the back of the eye, more light is used by the cat and less is wasted than if the reflecting layer weren't there.

Cats aren't the only animals that have reflective tissue in their eyes. Cattle, oxen, opossums, alligators, and some fish are among the other animals that have it. We humans don't, and that's part of the reason we're not as good as cats at finding our way in the dark.

Further Reading

McGraw-Hill Encyclopedia of Science and Technology, s.v. "Eye (Vertebrate)." New York: McGraw-Hill, 1987.

Minnaert, Marcel. *The Nature of Light and Colour in the Open Air*. New York: Dover, 1954.

Polyak, Stephen L. *The Vertebrate Visual System*. Chicago: University of Chicago Press, 1957.

The Shape of Lightning Bolts

As a thundercloud moves through the air, a strong negative charge gathers near its base. Because opposite charges attract, this negative charge is anxious to combine with the positive charges in the ground. Eventually a lightning bolt forms to neutralize these different charges.

We might think this bolt would want to jump in a straight line, that the electric charge would try to find the most direct route between thundercloud and ground. Why then are lightning bolts so jagged and irregular?

The answer has to do with the complex way a lightning bolt forms. Although it looks like it forms all at once, a lightning bolt is actually produced in many steps. Instead of jumping right to the ground, the cloud's negative charge begins with a short downward hop. This initial hop is called a leader, and it's no more than a few hundred feet long. From the lower end of this leader, another leader forms, and from the lower end of this, another. In this manner, the negative charge hops downward from leader to leader like a frog jumping from lily pad to lily pad across a pond. While this is going on, the ground sends up its own chain of shorter, positively charged leaders. It's only when these two chains meet, about a hundred feet off the ground, that we see the lightning bolt's flash.

So lightning is jagged because each leader forms independently of the others. Each place a lightning bolt zigs or zags is where one leader stopped and another one started. Each place a lightning bolt forks is where two separate leaders formed from the bottom end of a single leader above. This whole process takes only a few thousandths of a second, but that's enough time to sculpt beautiful and complex lightning bolts.

Further Reading

Trefil, James S. *Meditations at Sunset: A Scientist Looks at the Sky.* New York: Collier, 1987.

Alcohol in Pie . . . and Fried Fish?

If you're entering the county fair pie-baking competition, here's a tip: use vodka.

No, you won't make the judges tipsy. Instead, the special ingredient of ethanol—the chemical name for alcohol—makes a particularly light and flaky crust.

A pie crust is made of flour, some form of fat (such as butter or lard), and a liquid, which is often water. The liquid in your crust is necessary because it holds the dough together. But when you use water, you activate gluten. Gluten is a combination of two proteins that structure the dough as they're stretched in the mixing. While having some gluten in your crust is good, too much makes the dough tough.

Vodka that is 80 proof, at only 60 percent water, holds the flour and fat together without activating too much gluten. Use plain vodka as half the required liquid in your dough and most of the ethanol bakes off in the oven, leaving you with a flaky crust (and no vodka flavor). Or replace all the water with a strong spirit if you want to taste, say, a bourbon-infused crust with your pecan pie.

While at the fair, you might discover alcohol in other bready foods too, such as in the batter for fried fish. Beer contains carbon dioxide, which dissolves at low temperatures rather than hot. So when that beer batter hits the hot frying oil, it bubbles because it isn't soluble. Foaming agents in the beer, including some proteins, stop these carbon dioxide bubbles from bursting quickly. The result? A bubbly batter that's lacy and crispy.

Now you know: whatever crust you're making, add a little alcohol, and you can wow the judges.

Further Reading

Eplett, Layla. "How Alcohol Makes a Flakier Pie Crust: The 'Proof' Is in the Pie." *Scientific American*, March 14, 2014. https://blogs.scientificamerican.com/food-matters/how-alcohol-makes-a-flakier-pie-crust-the-proof-is-in-the-pie/.

Gibbs, Wayt W., and Nathan Myhrvold. "Beer Batter Is Better." *Scientific American*, February 1, 2011. https://www.scientificamerican.com/article/beer-batter-is-better/.

How Time Passes in Dreams

Many people believe that hours' worth of events and activities can be dreamed about in a matter of seconds. Despite this common belief about how we dream, time in dreams actually is not compressed. If you dream of an activity that would take five minutes in waking life, you probably dream about it for a full five minutes.

Dream and sleep researcher William Dement conducted two studies that demonstrated that dream time was similar to real time. Because dreamers' eyes move under their eyelids very rapidly while they are dreaming, Dement was able to monitor sleepers and record the length of their dreams by observing their rapid eye movement.

After recording this information, Dement would wake dreamers and have them write down a description of their most recent dream. He assumed that longer dreams would take more words to describe than shorter ones. When he compared the number of words in each dream report with the number of minutes over which the dream had occurred, he found that the longer the dream, the more words the dreamer used to describe it.

In another related experiment, Dement woke sleepers while they were dreaming and asked them how long they perceived their most recent dream had taken; 83 percent of the time they perceived correctly whether their dreams had been going on for a long time or for a short time. With these experiments, Dement concluded that time in dreams is nearly identical to time in waking life.

So the next time you slay a dragon or fly from your house to your workplace in your dreams, the amount of time it seems to take will probably be just about how long it actually takes to dream it.

Further Reading

Kelly, Dennis D. "Sleep and Dreaming." In *Principles of Neural Science*. 3rd ed. New York: Elsevier, 1991.

Why You Can Never Get to the End of the Rainbow

It seems that anything as beautiful as a rainbow must lead to a wonderful place where it touches the ground. But if you've ever tried to approach a rainbow, you know it recedes as you move toward it. The rainbow appears beyond that stand of trees or over that next hill. When you move, the rainbow moves with you.

Sooner or later you realize the end of the rainbow is not in any definite place you can mark on the map. No part of a rainbow is in any definite place, except in relation to your eye.

A rainbow is just a total of all the light coming to your eye from certain directions. The light from a rainbow is sunlight, reflected and broken into colors by water drops.

A rainbow always forms part of a circle, and the center of that circle is the point opposite the sun—from your point of view. The rule is that any water drops forty-two degrees of angle away from that point opposite the sun contribute to the rainbow you see.

Whether the point opposite water drops are ten feet away or ten miles away, they reflect light at that same angle and contribute to the same rainbow—for you. If you walk toward the end of the rainbow, it stays ahead of you as long as there are water drops in the air ahead of you.

Further Reading

Greenler, Robert. *Rainbows, Halos, and Glories*. New York: Cambridge University Press, 1980.

Minnaert, Marcel. *The Nature of Light and Colour in the Open Air*. New York: Dover, 1954.

Nussenzveig, H. Moysés. "The Theory of the Rainbow." *Scientific American* 236, no. 4 (April 1977): 116–27.

Do the Best Dogs Come
from the Pound?

The best dog for someone else may not be the best dog for you. But there's something to the idea that a mutt from the animal shelter can give you the best of everything, including good health and a good disposition. There's a principle of heredity at work.

A mutt is a dog whose parents are completely unrelated—they have no ancestor in common. Whatever undesirable genetic traits the parents may carry are likely to cancel each other out in the offspring.

On the other hand, if two parents are closely related—as brother and sister, for instance—or if they have any recent ancestor in common, each parent might have inherited a copy of the same bad gene from that common ancestor. If two dogs with the same bad gene now breed, some of the puppies will probably inherit two copies of that bad gene—and two copies are usually necessary for a genetic problem to show up. This can happen in puppy mills where incestuous matings are used to produce dogs in quantity for fast cash.

Responsible breeders breed only dogs whose ancestry they've checked out, looking for genetic problems. Responsible breeders also give their puppies individual care so the puppies learn to get along with people.

So some of the best dogs do come from the pound, and some of them come from top-notch breeders. Let the buyer beware of dogs from unscrupulous operators who use inbreeding for the sake of quantity, not quality.

Further Reading

The Monks of New Skete. *How to Be Your Dog's Best Friend.* Boston: Little, Brown, 1978.

Cooking with Wine

Say you have some friends visiting from out of town. You're preparing dinner, and one of them comments on the delicious aroma coming from the kitchen. Over the years you've kept it a secret, but she's a good buddy so you tell her it's your beef recipe marinating on the stovetop. Your secret is the red wine in the sauce.

When you cook with alcohol, it binds to the fat and water molecules alike. Some flavors—such as herbs—are fat soluble, meaning they dissolve best in fat; most other ingredients in the sauce are water soluble. This means that in the marinade, a cup of wine can dissolve some of the flavor compounds in the recipe and carry them into other cells more effectively than a water-based sauce. Plus, like high heat or salt, alcohol breaks down proteins in meat, a process known as "denaturing." So the wine in the marinade helps tenderize the beef.

You should be able to taste the wine in the dish even after it is cooked. That's because when you cook with wine—or any spirit—it doesn't all disappear. Alcohol's molecules are volatile: they evaporate rapidly, which is why we often smell it so intensely. And since alcohol boils at 173 degrees Fahrenheit versus water's boiling point of 212 degrees, the alcohol in your pot evaporates much sooner than water. In fact, reduction, dilution, and cooking time are some of the best ways to reduce the percentage of alcohol left in your dish. But it's a common misconception that it will burn off entirely, especially if you've added a whole cup of strongly flavored wine.

Further Reading

Faculty of Science, University of Copenhagen. "Practical Cooking Tips for Your Red Wine Sauce." *ScienceDaily*, June 6, 2017. https://www.sciencedaily.com/releases/2017/06/170606135751.htm.

Joachim, David, and Andrew Schloss. "Alcohol's Role in Cooking." *Fine Cooking* 104. Accessed July 7, 2019. http://www.finecooking.com/article/alcohols-role-in-cooking.

Weeks, Kevin D. "Spirited Cooking: Keep Some Liquor in the Kitchen." National Public Radio. January 19, 2010. https://www.npr.org/templates/story/story.php?storyId=122730434.

Listening Underwater

If you've ever been underwater at a pool when someone jumped in near you, you know the sound of the splash is clearly audible. But telling where the splash came from is another matter. Even though water does a much better job of conducting sound waves than air, that extra conductivity makes it harder, not easier, to tell where a sound comes from.

Above the surface of the water, we can tell whether a sound comes from the left or the right because it strikes one ear a little sooner and a little more loudly. The more distant ear gets a smaller dose of the sound a little bit later because it's farther away from the source and because it's shielded by the head. Even though we don't notice this difference consciously, our brains can detect it and use it to decide which direction the sound is coming from. But sound travels five times faster in water than it does in air. Traveling that fast, the sound is detected by both ears at almost exactly the same moment. That's one reason that a sound underwater seems to come from all directions at once.

The other reason is that underwater sound waves pass directly into your head, bypassing your ears altogether. That's because body tissues contain such a large amount of water. Try plugging your ears underwater and listening for another splash of someone jumping in. It will be just as loud as the last splash when your ears were not plugged. With sounds coming into every part of your head at almost exactly the same time, it's no wonder the brain has trouble deciding what direction the splash came from.

While you're in the pool, try this next demonstration with your friends. Duck your head underwater and listen to the conversation. If they talk loudly enough, you'll hear the vowels—*a*, *e*, *i*, *o*, and *u*—but no consonants. So the words won't make sense. Sound travels very well underwater, but some sounds have more trouble than others getting from the air into the water.

But why the vowels and not the consonants? Every spoken sound is actually a combination of different sounds, some low, some high. Even though we don't notice the different sounds, the way they're combined is what gives each spoken sound its own character. In general, consonants contain a lot more high-pitched sounds than vowels. Those are sounds made of faster, smaller sound waves. Compared to consonants, vowels contain mostly low pitches: in other words, they are made of larger, slower sound waves.

When the small sound waves hit the uneven surface of the water, they get scattered in all directions like ping-pong balls landing on a rough road. The much larger, lower-pitch waves aren't affected as much by the little water waves because they hit a much wider area on the water's surface. If we think of a small sound wave as a little ping-pong ball on a rough surface, a larger sound wave is more like a big basketball, which is less affected by little bumps on the road.

Unlike balls bouncing on a road, sound waves pass through the water, but like the basketball, the large waves come through with less distortion. That's one reason you'll hear the larger, slower sound waves of vowels but not the high-pitched, short waves of consonants when you listen to speech underwater.

Further Reading

Miller, Mary K. "Science in the Bathtub." *Exploring* (Winter 1993).

Schiffman, Harvey Richard. *Sensation and Perception: An Integrated Approach*. 2nd ed. New York: John Wiley and Sons, 1982.

Why Bells Are Made of Metal

Why are bells made of metal? Or, to ask the question another way, why is metal a good material for bells?

Try a simple home experiment. Hold an ordinary stainless-steel kitchen table knife loosely between two fingers and tap it sharply with another knife. It rings like a bell.

When you tap on the knife, it flexes very slightly away from the place you tapped. Then it springs back. But here's the important part: The knife doesn't just spring back to its original shape. It springs back with so much energy that it overshoots and flexes the other way. And, having flexed the other way, the knife springs back yet again—with enough energy to overshoot the original position again and flex the other way again.

This flexing back and forth happens hundreds of times per second, and, in a piece of the right kind of metal, it may go on for five or ten seconds or more. That's why metal is a good material for bells.

Each flex compresses the air near the knife. That's a sound wave. The rhythm of the flexing back and forth is regular, so the sound waves are regular, and we hear a musical tone. In other words, the metal knife vibrates and makes a sound. A bell is a piece of metal in another shape doing the same thing.

Metal isn't the only kind of material that will ring like a bell. Glass also will do it. Some kinds of wood will ring for a short time: for example, the rosewood in xylophones and marimbas. Since these different materials share the property of ringing when struck, there must be something similar about the way the atoms in each material are held together.

Further Reading

Bronowski, Jacob. *The Ascent of Man.* Boston: Little, Brown, 1974.

Cotterrill, Rodney. *The Cambridge Guide to the Material World.* New York: Cambridge University Press, 1985.

An Inverted Image

This is a simple visual experiment. To do it, you need two three-by-five index cards.

Take one of the cards and make a tiny triangle of holes in it with a straight pin. Make the three holes about a sixteenth of an inch apart. Take the other index card and poke just one pinhole in it.

Now take the card with the triangle of pinholes and hold it very close to your eye. Hold the other card, the one with the single pinhole, about four inches in front of the first one. Face a strong light and look through the three holes at the single hole. You will see your triangle of pinholes upside down.

The holes in the two cards cast three thin beams of light into your eye, making three dots on your retina. Under normal circumstances, your retina gets a focused, upside-down image projected by the lens of your eye. Your nerves and brain in effect turn that image over to make it right side up. But in this experiment, the card with the three holes is so close to your eye that the lens can't focus the three dots into an image. The rest of your visual system, operating as usual, inverts the triangle of light on your retina anyway, and you see the pinhole pattern upside down.

It's a very striking effect; try it.

Further Reading

Lynde, Carleton J. *Science Experiences with Ten-Cent Store Equipment*. 2nd ed. Scranton, PA: International Textbook, 1951.

The Elastic Ruler

Of all the home demonstrations described in this book, this might be the simplest. All you need is a ruler and a saucepan full of water.

Hold the ruler vertically and lower it into the water on the side of the pan farthest from you. Bring your eye down to a position just a little higher than water level and look at the ruler. The submerged part of the ruler looks shorter than the part above water. Notice that the effect becomes more pronounced as you lower your eye closer and closer to the level of the water surface.

Now try this: fix your eye just a little above water level and slowly raise the ruler out of the water. The ruler seems to stretch as it emerges from the water. And, of course, it works in reverse: dip the ruler back in and it seems to compress.

The physical principle at work here is that light is bent whenever it crosses at an angle from one transparent medium into another.

You can try to imagine following light rays as they leave the bottom of the ruler on their way to your eye. Light rays leaving the bottom of the ruler at a steep angle—say, upward at forty-five degrees—get bent toward the horizontal when they leave the water. The light rays then travel through the air at a shallow angle, closer to horizontal. Light rays reaching your eye at a shallow angle appear to come from just beneath the surface of the water.

Further Reading

Lynde, Carleton J. *Science Experiences with Ten-Cent Store Equipment*. Scranton, PA: International Textbook, 1951.

The Sweet Spot on a Baseball Bat

There's a right way and a wrong way to hit a ball with a bat. If you hit the ball with the wrong part of the bat, your hands get stung and the bat may break. If you hit the ball on the "sweet spot" of the bat, you get the feeling of a solid connection with the ball—no stinging, no vibration, and no broken bat.

Every bat has a sweet spot. The location along the length of the bat varies depending on the shape of the bat and on how you hold it. You can find the sweet spot by gripping the bat handle with one hand in the same place as when you're swinging. Take a hammer in the other hand and gently tap the bat at various places along its length. At some point you will feel almost no vibration when you tap. That's the sweet spot. That's the best place for the ball to meet the bat.

If you don't have a baseball bat handy, you can find the sweet spot on a pencil. Hold the pencil loosely between two fingers and tap it sharply against the edge of a hard table. Try tapping various spots along the length of the pencil, from the far end to very close to your fingers. You'll find one point where the tapping does not sting your fingers. You'll probably also notice a difference in the sound as you get closer to the sweet spot.

Any swinging object has a sweet spot, or "center of percussion," as engineers call it. A good hammer or axe is designed so that if you grip the handle at the proper place, the sweet spot is right where the tool strikes the nail or the wood. As a result, you get the optimum in power and control.

Further Reading

Eschbach, Ovid W. *Handbook of Engineering Fundamentals*. 2nd ed. New York: Wiley, 1969.

Ingard, U., and W. L. Kraushaar. *Introduction to Mechanics, Matter, and Waves*. Reading, MA: Addison-Wesley, 1960.

Kirkpatrick, Paul. "Batting the Ball." *American Journal of Physics* 31 (1963).

Why Kids Can Sleep through Just About Anything

Have you ever been to a party and seen a child sleeping happily on a couch, undisturbed by the adults talking and laughing in the room? How can kids sleep so soundly when exposed to noise and commotion? The answer lies in the difference between how adults sleep and how children sleep.

There are four different stages of sleep, and all sleepers cycle between these stages up to six times each night. The biggest difference between how adults and children sleep occurs in stage-four sleep, which is called slow-wave sleep, because in this stage your heart rate and your blood pressure decrease, your brain is less active than at any other time, and your dreams, if you have any, tend to be vague and abstract. Slow-wave sleep is also called deep sleep. Children spend far more time in this stage of sleep than adults.

During deep sleep, you sleep so heavily that you lose control of many of your muscles. Your mouth can drop open, and you might drool. You probably won't be awakened by noises and activity around you. For the most part, as people grow older, they spend less and less sleep time in deep sleep. By the age of sixty, most people will spend almost no time in this stage of sleep. Children, however, spend most of their sleep time in deep sleep, so those children sacked out on a couch at a party are probably not going to be disturbed, even by a bunch of adults standing around having a good time.

Cold Water at the Bottom of the Lake

It's a warm summer morning, and you're standing in a freshwater lake with water up to your neck. You notice that your feet are colder than your shoulders. Cold water sinks, of course. But how much colder could the water be on the very bottom of the lake?

It depends on the lake—where it is and how deep it is. But in a freshwater lake the coldest water at the bottom will not be colder than thirty-nine degrees Fahrenheit.

Fresh water is denser at a temperature of thirty-nine degrees than at any other temperature. Any thirty-nine-degree water in a lake will go to the bottom. Water that's warmer or colder than thirty-nine degrees will float on top and be exposed to the weather.

So the water in a lake divides into layers according to temperature except for two brief periods each year when the surface water temperature matches the deep water temperature. The temperatures match for a few days in spring as cold surface water is warmed up by the sun and in fall when warm surface water cools off with the approach of winter. During these special times, called the spring turnover and the fall turnover, all the water in the lake has the same temperature.

Then wind stirs the water, mixing oxygen and nutrients throughout the lake. For fish and other things living in a lake, these turnovers are among the year's biggest events.

Further Reading

Reid, George K. *Pond Life*. New York: Golden Press, 1987.

Smith, Robert Leo. *Ecology and Field Biology*. New York: Harper and Row, 1974.

Bad Grades and Biological Clocks

You might wonder why it's so hard for students to do well in college. Can't they just apply themselves and study hard? It seems simple enough, but it's really more complicated than that. In 2018, two American researchers published evidence that individual differences in daily biological rhythms influence academic performance.

First the researchers studied the daily activity patterns of fifteen thousand students by measuring the times at which they logged into the university's servers. They classified the students' daily activity patterns into three groups: morning larks, daytime finches, and night owls. This strategy of using computer logins allowed the researchers to study a really large number of students. And once the students had been classified according to daily activity cycle, the researchers could study how well the students in each group did in classes held at different times during the day.

The researchers found that when there was a significant mismatch between a student's daily activity pattern and their class schedule, their academic performance was significantly impacted. The effect was worst for night owls—students who were most active and alert late in the day. The phenomenon is called social jet lag because it is similar to the disorienting effect of flying across numerous time zones.

Can anything be done to prevent this social jet lag? There is no one schedule that works best for everybody. But the research shows that when individual students are able to schedule classes in accord with their daily activity cycles, they perform better in those classes.

Further Reading

Adamson, Allan. "Poor Grades Linked to Class Schedules That Do Not Match Students' Biological Clocks." *Tech Times*, March 20, 2018. http://www.techtimes.com/articles/224094/20180330/poor-grades-linked-to-class-schedules-that-do-not-match-students-biological-clocks.htm.

CTV News. "Planning Classes in Sync with Biological Clocks Could Boost Students' Grades." March 30, 2018. https://www.ctvnews.ca/health/planning-classes-in-sync-with-biological-clocks-could-boost-students-grades-1.3865551.

IFL Science. "Students Are Out of Sync with the School Day and It's Badly Affecting Their Grades, According to a New Study." Accessed June 13, 2019. http://www.iflscience.com/health-and-medicine/poor-college-grades-correlate-biological-clocks-being-out-sync-class-times/all/.

Neuroscience News. "Poor Grades Tied to Class Times That Don't Match Our Biological Clocks." March 30, 2018. http://neurosciencenews.com/grades-circadian-clock-871.

The Twin Within

Has someone ever said to you, "I saw someone the other day that looked just like you. In fact, I wondered if she were your twin"? You insisted that unless there was something your parents had not told you, you were sure you did not have a twin!

The fact is that even if you don't have a twin, you could still be a twin—you could be a chimera. Now, wait a second. A chimera is a mythological beast with the head of a lion, the body of a goat, and the tail of a dragon, right? In mythology, yes, but in biology a chimera is an organism made up of two distinct genetic lines.

This uncommon scenario happens when twin embryos fuse in the womb. The fusion results in what looks like a single embryo, but the genetic material from each twin remains separate. So what does that mean? Does a chimera have one arm from one twin and the other arm from the other twin? Well, it's something like that, although it's not that the arms would look different, exactly, or that the body's organs wouldn't be compatible or not work properly. But they do contain different and distinct sets of chromosomes. So if you're a chimera, your liver could be composed of cells with one set of chromosomes while your heart, say, consists of cells with an entirely different set.

What happens when the embryos that fuse are of different sexes? Then the chimera could be a hermaphrodite. A pretty mind-boggling concept, yes?

Further Reading

Wikipedia. "Chimera (Genetics)." Last edited June 30, 2019. https://en.wikipedia.org/wiki/Chimera_(genetics).

Limeys

The derogatory slang epithet *limey* is short for *lime-juicer*.

The original lime-juicers were British sailors of the 1800s who got lemon or lime juice with their food in order to prevent scurvy, a condition characterized by rotten gums, weak knees, and fatigue. During the late 1700s, about one-seventh of the sailors of the British navy were disabled by this disease.

A Scottish naval surgeon, James Lind, collected information about scurvy and learned it had often been cured by a diet of fresh fruits and vegetables—which the gruel-eating sailors certainly weren't getting. Lind understood that it was impractical to carry a lot of fresh produce on a ship in those days. But he experimented and found that just the juice from lemons, limes, or oranges could cure and prevent scurvy. Thus he recommended that sailors drink lemon juice at sea. The navy eventually took Lind's advice and put lemon juice aboard British ships starting in the 1790s. By the mid-1800s limes were cheaper than lemons, so lime juice was used instead. The British sailors became *lime-juicers*, then *limeys*.

Today we'd say that what the British sailors were getting from the fruit juice was vitamin C. For a while vitamin C was called *the antiscorbutic substance* because it prevented scurvy. The streamlined generic name *ascorbic acid* was invented in 1933.

Further Reading

Dictionary of Scientific Biography, s.v. "James Lind." New York: Charles Scribner's Sons, 1973.

Encyclopedia Britannica, 15th ed., s.v. "Biochemical Components of Organisms" and "Medicine." Chicago: Encyclopaedia Britannica Inc., 1986.

Partridge, Eric. *A Dictionary of Slang and Unconventional English*, edited by Paul Beale. London: Routledge and Kegan Paul, 1984.

The Secret Life of Hiccups

We've all experienced them at one time or another, often right after a big meal. A normal bout of hiccups usually lasts only a few minutes and may contain up to seventy individual "hics." Unlike coughing or sneezing, which can help clear your airways, hiccups seem to serve no beneficial function in the human body. What's the story behind these strange convulsions?

Two separate things happen to your body when you hiccup. The muscles in your diaphragm, which normally control your breathing, contract with a sudden jerk. This causes a sharp intake of breath. At the same time, your vocal cords contract to stop this breath, resulting in a loud "hic."

This is all caused by a misfire in the nerves that control your diaphragm. These nerves run from your neck to your chest, and any unusual pressure or irritation along this length can cause a misfire. Thus, hiccups are often triggered by overeating, gulping your food too quickly, or eating something too hot or too cold. Stress can also cause hiccups.

There are many folk remedies for hiccups, but none seems to work for everyone. Such remedies include holding your breath, breathing into a paper bag, or drinking a glass of water without breathing. It's possible that by depriving the diaphragm muscles of oxygen, these remedies force them to resume a more normal breathing pattern.

Other remedies include pulling your tongue, sucking a lemon, or having a friend startle you. What these remedies have in common is that they trick your nervous system with a diversion, perhaps shocking the nerves that control the diaphragm into normal behavior. No one knows exactly why these remedies sometimes work.

It's extremely rare, but severe cases of hiccups do occur. If you have persistent hiccups that simply refuse to go away, you should probably consult a physician.

Further Reading

"Hiccups That Don't Go Away." Berkeley Wellness: University of California, November 30, 2018. https://www.berkeley wellness.com/self-care/home-remedies/article/hiccups -dont-go-away.

"Hiccups." In *ABCs of the Human Body*, edited by Alma Guinness. New York: Reader's Digest Press, 1987.

How Does the World Look to a Bee?

To describe light in a general way, you need to specify at least three qualities: its brightness or intensity, its color, and its polarization.

Polarization is a quality our eyes don't detect. We have no everyday words to describe polarization, so we have to resort to a more or less scientific description of it.

If we think of light as a wave traveling through space—something like a ripple crossing a pond—we can think of polarization as describing the direction in which the wave vibrates. The vibration in a light wave is always perpendicular to the direction the wave is traveling. But the vibration of light can be up and down, sideways, or any combination of the two.

If the vibrations are in random directions, the light is said to be unpolarized; if all the vibrations are in the same direction, it's completely polarized. Intermediate amounts of polarization are most common.

To our eyes, polarization makes no difference. But it has been known for decades now that insects in general, and bees in particular, can detect the direction a light wave is vibrating in. Bees navigate by referring to the direction of the sun. But they don't have to see the sun directly; all they need is a clear view of a small piece of the sky. The blue glow of the sky is polarized, and the direction and amount of polarization are different in every part of the sky depending on where the sun is. A bee can tell where the sun is by looking at the polarization of any small piece of the sky.

So bees have a dimension to their vision that we lack. In addition to color and brightness, bees see polarization. What does that sensation feel like? How does the world look to a bee? We can only wonder.

Further Reading

Konnen, G. P. *Polarized Light in Nature*. New York: Cambridge University Press, 1985.

Minnaert, Marcel. *The Nature of Light and Colour in the Open Air*. New York: Dover, 1954.

Schmidt-Nielsen, Knut. *Animal Physiology: Adaptation and Environment*. 3rd ed. New York: Cambridge University Press, 1983.

Cottonmouth

Here's the scene. The audience is waiting with bated breath, you're just about to give a speech, and suddenly you notice your mouth is as dry as sawdust. You haven't said a word yet! Why did you get cottonmouth before you even began?

The dry mouth that for many people goes along with public speaking or performance is connected to the fight-or-flight response. When you're nervous or afraid, the nervous system slows down all the body processes that aren't necessary for your immediate survival and amps up the ones that just might save your life—that is, if you're about to be attacked by a predator like a tiger or a bear. As the fight-or-flight response kicks in, your heart rate goes up and more blood goes to your heart and major muscle groups, so you can make a quick getaway or grab a big stick.

When the fight-or-flight response is in full swing, digestion comes to a standstill, because who cares about breakfast when you're looking into a tiger's mouth? The salivary glands in your mouth are part of your digestive system, so they too go through a temporary slowdown, producing little or no saliva. That's what makes your mouth feel dry. You might feel butterflies in your stomach, too, as digestion comes to a halt.

Cottonmouth, a pounding heart, and butterflies in the stomach can be a hassle, but there are some advantages to activation of the fight-or-flight response. You may feel more alert and energetic than usual, lending your performance a memorable flair.

Further Reading

New Scientist. "The Last Word: Speaker's Throat." January 3, 1998. https://www.newscientist.com/article/mg15721158 –100-the-last-word/.

Schueller, Gretel H. "Thrill or Chill." *New Scientist*, April 29, 2000.

Why Mowing the Lawn
Doesn't Kill the Grass

If you cut down an oak tree, the stump dies. But if you cut grass, you don't hurt it at all. That's because new growth on an oak tree is at the tips of the branches; new growth in grass happens at ground level. Also, grasses, unlike other plants, can replenish their leaves.

A blade of grass is the end of a long, narrow leaf. If you trace a grass blade back to the stem on a tall grass plant, you see that the blade comes from a sheath wrapped around the stem. At the base of that sheath is a node, a place where the stems of some grasses have a slight bulge. Nodes are where new growth happens in a grass plant. A short grass plant has at least one node near ground level, out of reach of the lawn mower; a tall grass plant may have several more nodes farther up along the stem.

When a grass blade is cut off by a lawn mower or a grazing animal, some as-yet-unknown signal is sent down to the node, stimulating it to produce more leaf. Grazing animals take particular advantage of this; eating the grass causes more to grow in its place.

The capacity to add new material to old leaves is characteristic of grasses. Other plants generally don't have this ability. An oak tree grows new leaves every year, but it can't replace part of an existing leaf. An oak leaf grows to a certain mature size and stops. If part of an oak leaf is cut away, it doesn't grow back. The ultimate fate of an oak leaf is old age and death; leaves of grass remain youthful all summer long.

Further Reading

Curtis, Helena. *Biology.* 4th ed. New York: Worth, 1985.

Salisbury, Frank B., and Cleon W. Ross. *Plant Physiology.* 3rd ed. Belmont, CA: Wadsworth, 1985.

The Consequences of Smallness

Put a teaspoonful of granulated sugar in a glass of water and stir. The sugar dissolves in a few seconds. Do the same thing with a single lump of very hard candy, and two or three minutes later some of the candy will still be undissolved. The amount of sugar is about the same in both cases. The difference is that the small size of the grains of granulated sugar give the sugar and water more opportunity to interact.

Dissolving happens only where sugar meets water. The more square inches of sugar exposed to water, the faster the sugar dissolves. In general, thousands of small grains have more surface area than one big lump of the same volume. It's amazing how much surface area you can get by grinding a solid lump into a fine powder. One cubic inch of material, divided into particles a hundred-thousandth of an inch wide, has a surface area of several hundred square yards.

This has many implications—in biology, for instance. Large living things are made of many tiny cells. Each cell must constantly adjust its internal balance of water, nutrients, and waste products to function properly. These substances get into and out of cells through their surfaces. The more square inches of cell surface a living organism has, the faster it can adjust its chemical balance. So it's advantageous to be made of small cells.

Bibliography

Curtis, Helena. *Biology*. 4th ed. New York: Worth, 1983.

Kieffer, William A. *Chemistry: A Cultural Approach*. New York: Harper and Row, 1971.

Antimatter

Antimatter is not just the stuff of science fiction; it's real. Antimatter is made of particles complementary, in a way, to matter particles. For instance, the electron, a familiar matter particle, has a negative electrical charge. Its corresponding antiparticle, the positron, has a positive charge. When an electron and a positron meet, they annihilate each other. Both particles disappear in a flash of light—a spectacular case of matter changing to energy. The same thing happens in any encounter of an equal amount of matter and antimatter.

Antiparticles are created in nuclear reactions in stars and in space, but they soon meet matter particles and annihilate. So you're unlikely ever to see antimatter, even in a museum.

A great mystery about antimatter was discovered by the physicist Paul Dirac in 1930. Dirac found that, according to the known laws of physics, antimatter has just as much right to exist as matter. So why does so little antimatter exist today? One guess says that the very early universe—fifteen or twenty billion years ago—was made of almost exactly equal amounts of matter and antimatter. But, for some reason, there was just slightly more matter—by about one part in a billion. Annihilation eliminated most of the antimatter, but some unannihilated matter was left over to become our universe.

Whatever force or process it was that upset the symmetry between matter and antimatter in the early universe may still be at work today. That force or process may be the cause of unexpected behavior among subatomic particles like K mesons.

Clues about the whole universe in the distant past may be lurking among tiny particles in the present.

Further Reading

Adair, Robert K. "A Flaw in a Universal Mirror." *Scientific American* 258, no. 2 (February 1988): 50–56.

Feynman, Richard P. *The Feynman Lectures on Physics.* Reading, MA: Addison-Wesley, 1964.

How Dogs Eat

If you've ever watched a dog eat, you've probably marveled at how quickly it gulps down its food. You might even wonder why, whether a dog is hungry or not, it will often eat as much food as you put in front of it.

Dog owners may be concerned about this behavior, but it poses no problems for the dog. People chew their food and try to teach their children to eat slowly because digestion for humans begins in the mouth. Our saliva mixes with food and prepares that food to be broken down into its primary nutrients once it enters the stomach. A dog's digestion, on the other hand, doesn't begin until the food reaches the stomach, so dogs do not need to take time chewing their dinners.

Most dogs probably eat so quickly because in the days before they were domesticated, they had to survive by eating their prey before another dog or scavenger animal stole it. The evolutionary programming of dogs dictates that they eat and keep moving. As a species in the wild, they didn't have the luxury of hanging around and eating at their leisure.

Even their teeth aren't designed for them to savor their food. While most of the teeth in a human's mouth are flat and designed to facilitate chewing, most of the teeth in a dog's mouth are pointed and designed to allow the dog to grab its food and swallow it whole.

Hundreds of years of domestication haven't changed most dogs' eating habits very much. Even if a dog has been given regular, dependable meals every day, it will still gulp those meals down in a flash, ensuring that no scavenger will take its food away.

Further Reading

Johnson, Norman H. *The Complete Puppy and Dog Book*. New York: Atheneum, 1977.

The Secret of Clear Ice Cubes

The problem: your ice cubes are cloudy, with unsightly bubbles in the center, even though you started with clear water and a clean ice tray.

The answer: start with hot water, not cold.

The reason: hot water holds less dissolved air than cold water.

Those bubbles in the center of an ice cube come from air dissolved in the water. Bubbles usually form at the center because ice cubes usually freeze from the outside. The top, bottom, and sides of the cube freeze first, leaving a liquid water center. As the cube continues to freeze, dissolved air is forced into the liquid center. Air can't freeze at these temperatures, so when the liquid center of the ice cube finally freezes, the air comes out of solution and forms bubbles in the ice. Hot water has less dissolved air to begin with, so it makes fewer bubbles when it freezes.

To convince yourself that hot water holds less dissolved air than cold water, think of what happens when you heat water in a saucepan on the stove. Long before the water gets hot enough to boil, tiny bubbles form on the bottom of the pan. Those tiny bubbles are air coming out of solution as the water warms up. The same thing happens in your water heater.

Or think of an aquarium: if the temperature is too warm fish die, partly because the warm water holds too little oxygen.

Getting back to ice cubes, if some dissolved air has already been removed from water by heating, less air will be left to emerge as bubbles when you freeze the water in an ice tray. So the secret to making clear ice cubes is to start with hot water.

Further Reading

Delorenzo, Ronald A. *Problem Solving in General Chemistry.* Lexington, MA: D. C. Heath, 1981.

Broken Cups and Atoms

You can gather the pieces of a broken coffee cup and fit them together, but they won't stick. The fit may be good, but you can't make it good enough.

"Good enough" means getting the pieces so close that atoms interact.

Atoms separated by more than a few times their own diameter won't interact—they're basically indifferent to each other. The reason has to do with the inner structure of atoms, which is apparent only at very close range.

Every atom has a nucleus with a positive electrical charge surrounded by electrons with a negative electrical charge. Electrical opposites attract, so the nucleus and electrons attract each other.

Seen at a distance, though, an atom shows no obvious sign of positively and negatively charged parts. At a distance, electrical effects of the nucleus and electrons are canceled out. Only at close range—less than a few times the diameter of one atom—do the nucleus and electrons have distinct electrical effects.

In the same way, at a distance of a quarter mile, a red-and-whitecheck tablecloth shows no obvious sign of colored squares—it looks pink. Only at close range do the red and white squares look distinct.

If two atoms are brought close enough—in the ballpark of a ten-millionth of an inch—the nucleus of one may attract electrons of the other. The atoms interact, and a bond forms.

When a coffee cup breaks, the atoms are pulled apart so their relationship changes from interaction and bonding to indifference. The inner structure of an atom on one side is no longer apparent to an atom on the other side.

If you want the broken pieces of the cup to stick, you will have to fill the gap with something that will get close enough to the atoms on each piece to interact with them: glue.

Further Reading

Cotterill, Rodney. *The Cambridge Guide to the Material World.* New York: Cambridge University Press, 1985.

One-Way Glass

This Moment of Science is inspired by the plethora of crime dramas on television. It's going to unveil the magic behind one-way glass.

The trick is simpler than you might think. Most mirrors are made by applying a thin layer of a reflective material, aluminum in most cases, to the back of a sheet of glass. This is called back silvering, and it makes the glass opaque. When we look in a mirror, our image is reflected by the aluminum, which is made more durable by its glass covering.

The trick is that one-way glass is not fully silvered. The reflective material is applied less densely. This is called half-silvering. The effect is that the glass is not completely opaque like a traditional mirror. About half the light striking the glass passes through it, and the other half is reflected. It would seem like this would just result in people on both sides seeing the same thing: fractured images of both themselves and the people on the other side.

Now for the second trick to one-way glass: the lighting. The room the suspect is in is kept bright, so the reflective quality of the glass prevails. The room on the other side of the glass is kept dark, so instead of their reflections, the detectives see what is illuminated on the suspect's side of the glass: the suspect. If the light were to be turned up on the detective's side or down on the suspect's side, however, the magic would fizzle, and the glass would become a window for both parties.

Further Reading

HowStuffWorks. "How Do One-Way Mirrors Work?" Accessed June 14, 2019. http://science.howstuffworks.com/question 421.htm/printable.

McCarthy, Erin. "How Do Two-Way Mirrors Work?" *Mental Floss*, November 2, 2012. http://mentalfloss.com/article /12969/how-do-two-way-mirrors-work.

Wikipedia. "Mirror." Last updated June 6, 2019. http://en .wikipedia.org/wiki/Mirror.

WiseGEEK. "How Are Mirrors Made?" Accessed June 14, 2019. http://www.wisegeek.com/how-are-mirrors-made.htm.

Late Night Radio

Why do distant AM radio stations come in better at night?

The story begins in the rarefied atmosphere high above our heads, where ultraviolet light and X-rays from the sun strip electrons off atoms. The result is a gas made partly of electrons and partly of atoms from which electrons have been stripped. The stripped atoms are known as ions, and the part of the atmosphere where sunlight makes ions is known as the ionosphere.

When radio waves encounter the ionosphere, interesting things happen; ions and electrons have electrical charges that affect radio waves in many complex ways.

A radio wave of the right frequency coming up from the ground will be turned around and sent back down by the upper layers of the ionosphere a hundred or two hundred miles up. The upper ionosphere acts like a mirror, reflecting radio waves around the curve of the earth to distant receivers. That's why you can pick up a baseball game from a station hundreds of miles away.

But why is nighttime the right time for long-distance AM radio?

The reason is that during the day, the radio waves don't even get a chance to reach those reflecting layers of the ionosphere. The waves are blocked by a low-altitude layer about fifty miles up that exists only in the daytime. As long as the sun is up, this lower layer prevents AM radio signals from reaching the upper ionosphere. But the low-altitude layer of the ionosphere can't exist without continual sunlight. As soon as the sun goes down, the ions and electrons in this low layer get back together to form ordinary air. The obstruction disappears, and radio waves have a clear shot at the upper atmosphere, a hundred miles up in the night sky.

Further Reading

McGraw-Hill Encyclopedia of Science and Technology, s.v. "Radio," "Radio Broadcasting," and "Ionosphere." New York: McGraw-Hill, 1987.

Why Honey Turns Hard

If a little honey on a piece of homemade toast or in a cup of tea is how you like to start the day, you're not alone. Cave paintings show people a thousand years ago enjoying honey. One drawback to honey, though, is that after sitting too long on the shelf it crystallizes, and that soft, amber liquid turns to a hard, gooey mass.

Actually, though, only part of the honey is crystallizing. Honey is made mostly of two kinds of sugar: glucose and fructose. What crystallizes is the glucose, so the more glucose there is in comparison to fructose, the more likely it is to crystallize. Some honeys, like those made from the nectar of tupelo, locust, or sage, contain slightly more fructose than glucose, so they crystallize more slowly.

But before honey can crystallize, it needs what's called a "seed" for the crystals to grow on. The seed might be a grain of pollen, a speck of dust, or even a scratch on the inside of the jar. But the best seed of all is a bit of honey that has already crystallized. Most of the honey in a supermarket has been heated and filtered to remove virtually all the possible seeds. That slows the crystallization, but the heating process also drives off some of the honey's distinctive flavor. When honey does crystallize, you can soften it again in a microwave or a pan of warm water, but as it cools, the crystallization will begin again—even faster than before.

Honey crystallizes faster the second time because heat alone can't remove all the seeds. Dust, crumbs, and other tiny particles that have accumulated since you first opened the jar will remain as seeds to start the process all over again.

Further Reading

McGee, Harold. *On Food and Cooking: The Science and Lore of the Kitchen*. New York: Scribner, 1984.

Adding and Subtracting Colors

Red plus blue plus green makes white when you mix light and black when you mix paint. Mixing colored light and mixing colored paint are different processes.

You can see mixing of colored light in a color television picture. All the colors we see in the picture from a distance are made of glowing dots of red, green, and blue light on the screen. These dots glow with different intensities to make different colors. If only the red dots glow, the TV picture is red. If the red dots and the green dots glow, the picture looks yellow because red light added to green light makes yellow light. If all the colored dots glow at full strength, the picture looks white. Other colors are made by adding the three basic colors in various proportions.

Paint, on the other hand, doesn't make light. It absorbs light. Take the example of a red car parked in the sun. Sunlight already contains all the colors of the rainbow. The red paint on the car absorbs all non-red colors. In other words, red paint subtracts just about every color from white light, so red looks closer to black than white.

When you mix paint, you're subtracting colors. When you mix light, you're adding colors. So you have to think of both adding and subtracting color when you try to guess, for instance, the color of a light-blue wall illuminated by a yellow-orange light bulb.

Encyclopedia

McGraw-Hill Encyclopedia of Science and Technology, s.v. "Color Television" and related entries. New York: McGraw-Hill, 1987.

Breaking a Coffee Cup

Why is it so easy to break a coffee cup if it's already cracked?

To break a coffee cup, you have to pull atoms far enough apart—maybe a millionth of an inch—that they no longer bond to each other.

All along the crack, atoms have already been separated. The point of interest is at the tip of the crack—where it ends in solid material. To make the crack just a little longer, all you have to do is separate a few more atoms.

The important principle here is leverage.

Leverage is the principle that allows you to pull a nail out of hard wood with a crowbar. A small force at the end of a long crowbar becomes an immense force pulling on the nail. The crowbar has the effect of magnifying the force you apply to it.

Usually, a cracked coffee cup breaks because something hits it or presses on it with a force. The two sides of the crack act as levers, transmitting that force to the tip of the crack like little crowbars, pulling atoms apart. As the crack proceeds through the cup, these levers get longer, so they pull atoms apart with even greater force. The longer the crack becomes, the easier it is to make it just a little longer.

To break a coffee cup, you don't really have to pull billions and billions of atoms apart all at once—you only have to separate a few atoms at the tip of the crack, then a few more after that, and a few more after that, and so on all the way to the other side of the cup.

In real life all this happens in a split second as the cup hits the floor. High-speed photographs show that cracks travel through brittle materials like ceramics and glass at thousands of miles per hour.

A split second is all it takes to break a coffee cup.

Further Reading

Field, J. E. "Fracture of Solids." *The Physics Teacher* 2, no. 5 (May 1964). https://doi.org/10.1119/1.2350789.

Gilman, J. J. "Fracture in Solids." *Scientific American* 202, no. 2 (February 1960): 94–107.

Déjà Vu

"Have you really been there before?"

Many people at one time or another have experienced déjà vu. French for "already seen," déjà vu is a sudden strong feeling that a moment identical to the present one has occurred at some earlier time.

To a cognitive psychologist, déjà vu is proof of the immense amount of knowledge and experience we store in our brains.

When we experience déjà vu, what actually happens is that, in a fraction of a second, we retrieve bits of many different memory fragments and piece them together, producing what seems to be a complete memory. So, if you experience déjà vu in a mall restaurant while waiting for a pepperoni pizza with your best friend, your mind has taken perhaps hundreds of stored memories of various experiences and put together fragments from those memories to give you the sensation of having been there before, even though you haven't been there before at all.

Cognitive psychologists who study how we use language are not surprised at the brain's ability to create déjà vu. Actually, language comprehension and déjà vu have many parallels. When you hear someone speak, you usually understand her even though you've probably never heard her words presented in exactly the same way. You understand these sentences because your brain is able to remember the individual meanings of words based on hundreds of past experiences with those words. Your brain takes the meanings of individual words and splices them together to comprehend their meaning as a whole. As with déjà vu, this entire process happens in a split second.

A Rock in a Row Boat

Here's a classic puzzle you'll enjoy pondering.

You're in a rowboat in a small pond. You have a huge boulder with you in the boat. You throw the boulder overboard into the water. Does the water level of the pond rise, fall, or stay the same?

Now, it's pretty clear that if you took everything completely out of the pond—yourself, the boat, and the boulder—the water level would fall. And if you stayed in the boat, in the water, and threw the boulder ashore, the water level in the pond would also fall. Since the rowboat would weigh less after you tossed the boulder ashore, the boat would float higher and displace less water. But what happens if you toss the boulder overboard into the water?

You can simulate the situation by filling a big mayonnaise jar with water. That's the pond. Put a rock in a tin can—that's the rowboat—and float it in the water in the jar. Mark the level of water in the mayonnaise jar. Next take the rock out of the can and drop it into the jar. Now you have a rock at the bottom and the empty tin can floating at the surface. Don't let it capsize. Is the level of water in the jar higher, lower, or the same as before?

Here's the logic behind the answer: When the boulder is in the boat, it displaces an amount of water equal to its weight. That's a lot of water. Lying at the bottom of the pond, the boulder displaces only an amount of water equal to its volume. That's less water. The bolder displaces less water if it's on the bottom of the pond than if it's in the boat. Also, the boat floats higher without the rock.

So here's the answer: if you're in a rowboat with a big boulder and you throw the boulder overboard into the water, the pond water level falls.

Further Reading

Brown, R. J. *333 More Science Tricks and Experiments*. Blue Ridge Summit, PA: Tab, 1981.

When It Smells Like Rain

If you've been outside before a rainfall in the spring, you might have noticed a particularly fresh or sweet smell that seeps into the air a few minutes before the first drops begin to fall. If you're familiar with this smell, it's a great way to predict when rain is about to start. What, exactly, makes the air smell like rain?

A lot of stories and folktales have arisen to explain this odor. Long ago, people used to believe that rain clouds picked up sweet smells from heaven and the rain carried these angelic odors to Earth.

In fact, that fresh smell isn't coming down from the sky at all. It's coming up from the ground beneath your feet.

As a spring rain approaches, the humidity level near the ground tends to increase. Moist air is much better than dry air at transmitting smells—this is why you might use warm, steaming water to carry the smell from dried potpourri into the air of your house. As the humidity rises, the moist air carries the fresh smell of oils, secreted by grass and other ground plants, up to your nose. It also carries the odors of ground-dwelling bacteria and fungi.

These smells are always in the air immediately above the ground, as you can test for yourself by sticking your nose into your lawn and breathing deeply. When a rainstorm approaches and the humidity climbs, these odors rise up from their earthly embrace, and you say it smells like rain!

Further Reading

Dennis, Jerry. *It's Raining Frogs and Fishes*. New York: Harper Collins, 1992.

Stromberg, Joseph. "What Makes Rain Smell So Good?" *Smithsonian*, April 2, 2013. https://www.smithsonianmag.com /science-nature/what-makes-rain-smell-so-good -13806085/.

Wikipedia. "Petrichor." Last edited April 27, 2019. https://en .wikipedia.org/wiki/Petrichor.

Yuhas, Daisy. "Storm Scents: It's True, You Can Smell Oncoming Summer Rain." *Scientific American*, July 18, 2012. https://www.scientificamerican.com/article/storm-scents -smell-rain/.

Mirages

On a hot, sunny day, the road in the distance seems to be covered with water. It reflects the blue of the sky. But when you get there, the mirage is gone.

You realize, of course, that there was never any water there in the first place. But the apparition was convincing; what you saw looked exactly like light reflected from a puddle. And as far as the reflection of light is concerned, there might as well have been water on the road.

The pavement, baking in the sun, is hot. The hot pavement heats the air near the surface to a much hotter temperature than the air higher up. A few inches above the pavement there's a boundary between hot air and cooler air—a boundary that can reflect light just like a water surface.

Light can be reflected from any boundary between one transparent medium and another. Take, for example, the boundary between air and water. If you look toward the far side of a quiet lake, you see trees and mountains beautifully reflected from the top of the water surface.

Light can be reflected from the bottom of a water surface, too. If you're underwater in a swimming pool, wearing goggles so things are in focus, look up and toward the far side of the pool. If the water is quiet, you will see a mirror-like reflection from the underside of the water surface.

Light can be reflected from either side of a boundary between one transparent medium and another. The reflection is best if light hits the boundary at a glancing angle.

Back to the road: The boundary between hot air near the pavement and cooler air a few inches up reflects light exactly as a water surface would. From a distance, viewing the road at a glancing angle, it's hard to tell whether you're seeing water or just reflection from the top of a layer of hot air—a mirage.

Further Reading

Minnaert, Marcel. *The Nature of Light and Colour in the Open Air.* New York: Dover, 1954.

Why Popcorn Pops

Popcorn, like all grains, contains water. About 13.5 to 14 percent of each kernel is made up of water. So when a popcorn kernel is heated above the boiling point of 212 degrees Fahrenheit, this water turns into steam. The steam creates pressure within the kernel, causing the kernel to explode and turn itself inside out. But if the water inside a piece of popcorn is what makes it pop, why don't other grains pop as well? Wheat and rice contain water, so why don't we sit down to watch a movie with a bucket of popped rice or popped wheat?

The answer lies in the differences between the outer coverings, called hulls, of popcorn and other grains. Unlike rice and wheat, and unlike even regular corn, popcorn has a nonporous hull that traps steam. With the porous hulls of other grains, steam easily passes through, so no significant pressure is produced. These grains may parch, but they will not pop.

But even popcorn, with its special hull, doesn't always pop. Popcorn must have two important properties to pop well. First, the amount of moisture in the kernel must be very close to 13.5 percent. Too little moisture, and not enough steam will build up to pop the kernel. Too much moisture, and the kernels pop into dense spheres, rather than the light, fluffy stuff popcorn fanciers love.

Second, the kernels must not be cracked or damaged in any way. Even a small crack will let steam escape, keeping the necessary pressure from building. Popcorn kernels with the right amount of moisture and unblemished hulls pop into the snack that just about everyone enjoys.

Make an Image without a Lens

Punch a pinhole in the center of a big piece of cardboard. That pinhole can project an upside-down image onto a piece of white paper.

The effect is easiest to see if you're indoors on a sunny day in a room with a window. Stand on the side of the room farthest from the window. Hold the card with the pinhole vertically. Hold a piece of white paper in the shadow of the card, behind the pinhole. Light coming through the pinhole will make an upside-down image of the window on the white paper.

Hold the white paper closer to the pinhole to get a brighter but smaller image. Make the pinhole bigger by shoving a pencil point into it to get a brighter but fuzzier image.

A way to see how it works is to think of rays of light traveling in straight lines from each part of the window, through the pinhole, onto the white paper. Because the pinhole is small, each point on the white paper receives light from only a tiny part of the window. So light from the window is formed into an image on the white paper.

You can also see that light from the top of the window, after going through the hole, ends up at the bottom of the projected image. That's why the image is upside down.

Making an image with a pinhole is so simple any child can do it. But the technique is also used on the cutting edge of modern science. Some objects in space, including supernovas and black holes, emit X-rays. Ordinary lenses and mirrors cannot focus X-rays, but a pinhole can. Many X-ray telescopes aboard astronomical satellites are elaborate pinhole cameras based on the same principle you've just demonstrated with a piece of cardboard.

Further Reading

Skinner, G. K. "X-ray Imaging with Coded Masks." *Scientific American* 259, no. 2 (August 1988). https://www.scientific american.com/article/x-ray-imaging-with-coded-masks/.

Lynde, Carleton J. *Science Experiences with Ten-Cent Store Equipment*. 2nd ed. Scranton, PA: International Textbook Co., 1951.

A Rising Fastball

Suppose a baseball pitcher releases the ball with a sharp downward snap of the wrist. That snap of the wrist imparts backspin to the ball. Backspin is opposite in direction to the spin of a ball just rolling on the ground. If the ball has backspin, the ball's top surface moves in the same direction as the flow of air over the ball and the bottom surface moves in the opposite direction to the flow of air. So, because of friction between the surface and the air, the top surface of a ball with backspin helps the air along, so to speak. And the bottom surface slows the air down.

Now, a famous principle: A fast-moving airstream has lower pressure than slow-moving or still air around it—a principle named after the eighteenth-century physicist Daniel Bernoulli, who discovered it.

From the point of view of a ball with backspin, the air on top moves faster and therefore has lower pressure than the air on the bottom. The result is a force pushing the ball up. A baseball thrown hard enough, with enough backspin, *will* rise—or at least seem to be suspended—as it approaches the plate. That's your basic fastball.

This effect was first described in 1852 by the German physicist Heinrich Magnus. He was thinking about spinning cannonballs rather than spinning baseballs, but the principle is the same: Spin makes air flow faster along one side of the ball than the other. That makes a difference in air pressure, which in turn influences the flight path.

Further Reading

Gray, H. J., and Alan Isaacs, eds. *A New Dictionary of Physics*. London: Longman, 1975.

Oxford English Dictionary, 2nd ed., s.v. "Magnus Effect." New York: Oxford University Press, 1989.

Prandtl, Ludwig. *Essentials of Fluid Dynamics*. London: Blackie, 1952.

Chimes for Your Ears Only

This kitchen demonstration calls for a fork and a piece of string about five feet long. Tie the middle of the string tightly around the fork at the narrowest part of the handle. Hold one end of the string in each hand and let the fork hang down in the middle. Press the ends of the string to your ears and swing the hanging fork against the edge of a table so it strikes once and bounces away. If the fork you're using is made of one piece of metal, with no joints or rivets, it will ring like a bell and send a surprisingly loud and rich chiming sound through the string to your ears. If you try the same thing without pressing the string against your ears, the sound isn't nearly as rich. The reason for the loudness is that the taut string carries the fork's vibrations much better than air does. Take the string away from your ears, and all you hear is whatever comes to your eardrums through the air.

The reason for the richness of the sound is that you've suspended the fork in a way that allows it to vibrate freely, at many frequencies, making many musical tones at the same time. If you were holding the fork in your hand, the flesh on your fingers would dampen a lot of those vibrations.

Try tying not just one but two forks, or a fork and a spoon, into the middle of the string. Allow the cutlery to hang freely, press the ends of the string to your ears, and let the cutlery strike the edge of a table. From plain, ordinary flatware you can hear a sound almost like church bells.

Bibliography

Gardner, Martin. *Entertaining Science Experiments with Everyday Objects.* New York: Dover, 1981.

Lynde, Carleton J. *Science Experiences with Ten-Cent Store Equipment.* 2nd ed. Scranton, PA: International Textbook Co., 1951.

How Can You Tell If a Spider Is Dead?

If a spider is not moving and all its legs are flexed—that is, pulled in toward its body—it's likely to be dead. Although spiders' legs have flexor muscles—muscles that bend the legs in toward the body—they do not have extensor muscles—muscles that would cause the legs to straighten and point away from the spider's body.

So a spider flexes its legs by using its flexor muscles. But how does a spider extend its legs? In the 1940s the zoologist C. H. Ellis noted that, as a rule, dead spiders have flexed legs. Evidently, whatever straightens a spider's legs in life is inoperative in death.

Ellis and other zoologists demonstrated that spiders extend their legs with a hydraulic system. The legs of a living spider contain fluid under pressure that tends to straighten the legs, just like water pressure stiffening a garden hose or hydraulic fluid pressure lifting a car at a garage. The spider increases fluid pressure when it wants to extend its legs more forcefully. If a spider's leg is cut, the spider can't straighten that leg until it seals off the fluid leak.

If the spider dies, it can't maintain its internal fluid pressure. The leg flexor muscles may contract one more time, but without fluid pressure there will be no opposing force to straighten the legs again. That's why a motionless spider with flexed legs is likely to be dead.

Further Reading

Vogel, Steven. *Life's Devices: The Physical World of Animals and Plants*. Princeton, NJ: Princeton University Press, 1988.

Why Fan Blades Stay Dirty

Turn off your window fan for a moment, and you will see fine dust on the fan blades. It seems strange—when the fan is running, there's air blowing over the blades hour after hour. Yet that wind doesn't carry the dust away.

When you look at dirt on fan blades, you're looking at a manifestation of one of the most complex processes in nature: air flowing over a solid surface. Figuring out the details requires the fastest supercomputers in existence. But the overall picture is something like this:

If you could get small enough to sit on a spinning fan blade, you'd feel a strong breeze. It would be like riding in a convertible with the top down. But here's the strange and mysterious part: As you got still smaller and closer to the surface, you'd feel less and less breeze. Within a fraction of a millimeter from the fan blade surface, you'd feel no breeze at all!

Because of friction between air and fan blade, there's a very thin layer of air next to the surface that doesn't move over the blade. Any dust particles small enough to stay within that quiet surface layer never feel the breeze, so they stay put.

This fact has other practical implications. Blowing on a phonograph record won't get rid of the very smallest dust particles, especially the ones at the bottom of the grooves. And blowing on a camera lens will never get it really clean. In both cases you have to use a brush that touches the surface to get the fine dust off.

In his autobiography, *Slide Rule*, novelist and aeronautical engineer Nevil Shute Norway describes calm air near the surface of a 750-foot-long airship built in 1930: "When the ship was cruising at about sixty miles an hour, as soon as you got to the top, or horizontal, part of the hull you were in calm air crawling on your hands and knees; if you knelt up you felt a breeze on your head and shoulders. If you stood up the wind was strong. It was pleasant up there sitting by the fins on a fine sunny day" (101).

Bibliography

Ahrens, C. Donald. *Meteorology Today.* 2nd ed. St. Paul, MN: West, 1985.

Feynman, Richard P. *The Feynman Lectures on Physics.* Reading, MA: Addison-Wesley, 1964.

Khurana, Anil. "Numerical Simulations Reveal Fluid Flows near Solid Boundaries." *Physics Today* 41, no. 5 (May 1988).

Norway, Nevil Shute. *Slide Rule.* Portsmouth, NH: Heinemann, 1954.

Peterson, Ivars. "Friction Features." *Science News* 133, no. 18 (April 30, 1988): 282. https://doi.org/10.2307/3972603.

Peterson, Ivars. "Reaching for the Supercomputing Moon." *Science News* 133, no. 11 (March 12, 1988): 172–3. https://doi.org/10.2307/3972489.

Weisburd, S. "Record Speedups for Parallel Processing." *Science News* 133, no. 12 (March 19, 1988): 180. https://doi.org/10.2307/3972495.

The Legacy of the Dodo

The last dodo bird died in the late 1600s, probably as food for sailors. On the island of Mauritius in the Indian Ocean, where they lived, the large, flightless dodoes had been shot and eaten for years by sailors. Today we think of the dodo as the classic example of a peculiar anachronism, but in its own habitat the dodo played a critical role.

Also on the island of Mauritius was the Calvaria Major tree, which had evolved seeds so hard that they couldn't germinate by themselves. Instead, the seeds had to be cracked somehow before they could grow into young trees. Dodoes ate these seeds and digested the outer layer. By cracking the seeds and removing their outer layer, the dodo's gizzard also prepared the seeds for sprouting, and when they left the bird's body, they were ready to grow. When the last dodo died, there were no animals left on the island that could perform this necessary service for the Calvaria seeds. Today there are only about a dozen of these trees left on the island, all more than three hundred years old.

The relationship between the dodo and the Calvaria is one example of what biologists call coevolution—when two or more species evolve in ways that make them mutually dependent on each other. The dodo depended on the Calvaria for food, and the Calvaria depended on the dodo to make its seeds viable. Coevolution means that all species are part of a complex web, and for every species that goes extinct, several more extinctions may result. Sometimes species adapt to life without their coevolved partners, but often, as in the case of the Calvaria, the dependent species die off as well.

Further Reading

Temple, Stanley A. "Plant-Animal Mutualism: Coevolution with Dodo Leads to Near Extinction of Plant." *Science* 197, no. 4603 (August 26, 1977): 885–6. https://doi.org/10.1126/science.197.4306.885.

Get Your Bearings with Two Thumbtacks

Here's a way to learn a lot about your surroundings from one of those U.S. Geological Survey quadrant maps. They cover just a few square miles each in tremendous detail. You can get one for your area from the U.S. Government Printing Office or, possibly, from a local sporting goods store.

Lay the map out flat on a table near a window. Find your location. Shove a thumbtack through that point on the map into the table below. Now the map is free to rotate around the thumbtack, but the point corresponding to your location stays fixed.

Next, look out the window and find some prominent landmark, preferably a few miles away—a hilltop, for example, or a fire tower. Find the symbol for that landmark on your map. Rotate the map around the thumbtack until a straight line drawn from the thumbtack through the map symbol points to the landmark. Stick one more thumbtack through that map symbol into the table. You've now oriented the map using two known points: your location and a landmark you can see from your location.

Now that the direction to one landmark is correct, the directions to every other landmark will also be correct. A line of sight from the central thumbtack to, say, a hill whose name you don't know will pass through the symbol for that hill on your map. Look at the map, and you may find a name for the hill. Of course, if there's more than one hill along that line of sight, you may have to estimate the distance to the one you're looking at.

But you can get your bearings with two thumbtacks. Tack down your location first, then orient the map by referring to a landmark you can see. This works best with maps showing an area so small that the curvature of the earth is not noticeable.

Bibliography

Davis, P. J., and W. G. Chinn. *3.1416 and All That*. Boston: Birkhauser, 1985.

More Than One Way to Make a Frog

Most frogs in North America lay eggs in water. The eggs hatch into tadpoles, which have tails for swimming and gills to gather oxygen from the water. Eventually the tail and the gills disappear as the tadpole develops into an adult frog. But some frog species, especially in Central and South America, have a different life history: they skip the tadpole stage. In many tropical species, the eggs are laid in burrows in the ground, not in the water. The newly hatched frogs already look something like miniature adults. They never develop gills or even the circulatory system that would supply blood to gills. This so-called direct development, without a tadpole stage, seems to be well suited to tropical environments, where tadpoles swimming in open water would be in greater danger of being eaten than in colder climates like those in North America.

So different types of frogs have different ways of developing from embryos into adults. The difference in life histories is due not so much to a difference in basic body plan but to a difference in timing of development. For example, one type of tropical frog that skips the tadpole stage begins to develop legs very early—long before it hatches. In North American frogs, on the other hand, the development of legs is delayed until long after hatching: tadpoles have no legs. But the end results—the legs of the adult frog—are basically the same in tropical frogs and in North American frogs. This is a case in which the schedule of development from egg to adult has been modified by evolution to suit different environments.

Further Reading

Raff, Rudolf A., and Thomas C. Kaufman. *Embryos, Genes, and Evolution*. New York: Macmillan, 1983.

A Dot, a Line, a Crease, a Beautiful Curve

Take a piece of paper and a pencil. Draw a straight line somewhere on the paper—it doesn't matter where. Then draw one dot somewhere else on the paper—again, it doesn't matter where.

Now fold the paper over so that dot comes down somewhere on the line. Hold the dot on the line and crease the paper at the fold.

Open the paper up and refold it at a different angle. Put that dot somewhere else on the line, hold it there, and crease the paper at the fold.

Do this a few more times, always folding the paper at a different angle, but always putting that original dot somewhere on the line before you crease the paper.

Flatten the paper out, and you will see that the creases form a curved boundary around the dot. The more creases, the smoother the curve. You can trace through the curve with a pencil.

Technically speaking, that curve is a parabola, one of the most famous curves of mathematics and physics. The path of a stream of water squirting out of a garden hose is close to a parabola; so is the path of a ball thrown through the air. Air resistance and the curvature of the earth cause slight deviations from the parabolic ideal.

In space, where there's no air resistance, high-speed comets follow parabolic paths around the sun. The dot on your paper marks the position of the sun in the parabolic orbit. Make the dot and the line farther apart, and you'll get a close-up of the portion of the comet's orbit nearest the sun.

Further Reading

Gardner, Martin. "The Abstract Parabola Fits the Concrete World." Mathematical Games. *Scientific American* 245, no. 2 (August 1981): 16–28.

The Shape of Sound

For this kitchen demonstration you will need a candle, a piece of paper, and a big knife. Roll the paper around the candle, then put the candle on a cutting board and cut it in two with the knife. But cut it diagonally, like a green bean—in other words, cut straight down, but with the knife at an acute angle to the candle, not perpendicular to it.

Now unroll the paper and look at its cut edge. It's shaped in an undulating curve like the profile of ripples on a pond. That's one of the most important curves of mathematics and physics, the so-called sine curve.

Sine curves describe the nature of sound. The wavy edge of your paper could represent air pressure in a series of sound waves. Where the curve goes up, the air pressure is slightly higher than average; where the curve goes down, pressure is lower. When regular alternations of high and low pressure strike your eardrum, you hear a musical tone.

Look at the wavy edge of your paper and notice how far apart the waves are—the wavelength, as a physicist or engineer would call it. The distance between waves corresponds to the musical pitch of the sound. If the waves are closer together—in other words, if the wavelength is shorter—you hear a higher pitch. You can get a shorter wavelength by wrapping paper around a thinner candle and cutting it as you did before.

Look at the wavy edge and notice how far up and down the waves go. That corresponds to the loudness of the sound wave. You can vary that by changing the angle of the knife when you cut through the candle. If the knife is more nearly parallel to the candle, you'll get a sine curve with higher highs and lower lows.

Further Reading

Steinhaus, Hugo. *Mathematical Snapshots*. New York: Oxford University Press, 1950.

Blinking

On average we blink about 14,440 times a day. Since each blink takes about a quarter of a second, about an hour of our waking hours is spent with our eyes partly or completely closed. It is only in recent years that scientists have really begun to understand blinking. Probably most of us think the main purpose of blinking is to clean and lubricate our eyes, but it's not. We blink about fifteen times a minute, but only one or two of those blinks are needed to keep the surfaces of our eyes rinsed and moistened. We also blink a few times a day because of dust or smoke in our eyes or being startled. But for the most part, blinking is an indication of what's going on in our brains.

Generally speaking, the harder we concentrate, the less we blink. Car drivers blink less in city traffic than on the open road; they probably won't blink at all while passing a truck at high speed.

Further insights into blinking have been gained from studying subjects who were reading. They blinked most often at a punctuation mark or the end of a page. It was sort of like a signal that the brain was taking a break, a mental punctuation mark, as it were.

The concentration connection wasn't evident just in visual activities. Someone who is anxious tends to blink more than one who is calm. A steady gaze is associated with confidence and self-assuredness; TV anchors are instructed not to blink much to give the impression of being in control.

If blinking is a sort of mental punctuation, this might explain why people solving mental arithmetic blink at different rates.

Some do not blink until they have mentally solved the problem while others blink with each step in the process of solving it. If you want to test this, watch people on the sly so they don't become conscious of blinking, and although you won't be able to see what they are thinking, you might get a glimpse into *how* they are thinking.

Further Reading

Ingram, Jay. *The Science of Everyday Life*. New York: Penguin Books, 1989.

Stern, John A. "What's behind Blinking: The Mind's Way of Punctuating Thought." The Sciences. The New York Academy of Sciences (November–December 1988). https://doi .org/10.1002/j.2326-1951.1988.tb03056.x.

Sorting Out Musical Pitches

Sing into the strings of a piano with the damper pedal held down, and you will hear the reverberation of the notes you just sang. Pressing the pedal lifts the dampers from the strings, leaving the strings free to vibrate. The sound of your voice then causes the strings to vibrate, but not all equally. The strings that vibrate most energetically, and that keep vibrating after you stop singing, are the ones tuned to the pitches you sang.

Something a little like that happens in the human ear. Curled up in a fluid-filled capsule in the inner ear is a piece of tissue about an inch and a half long, the so-called basilar membrane, which is free to move in response to vibrations that come to it from outside. Just as different strings in the piano vibrate at different frequencies and make different musical pitches, different parts of the basilar membrane vibrate most energetically at different frequencies.

But there's a difference between the basilar membrane and a piano. The basilar membrane is one solid piece of elastic tissue, precisely shaped—thick at one end and thin at the other, like a tiny chisel. The basilar membrane does not have tuned strings; it has this special shape enabling it to sort out pitches.

The most energetic vibration is at the thick end for high frequencies, and at the thin end for low frequencies. Nerves connected to the membrane send signals to the brain indicating which part is vibrating most energetically.

If two or more musical notes sound together, the basilar membrane will vibrate strongly in two or more different places. That's why we can pick out individual notes in a musical chord.

The basilar membrane of the inner ear, because of its special shape, sorts out vibrations by frequency.

The highest pitches humans can hear correspond to about twenty thousand vibrations per second. How do we do that? Not by counting vibrations.

A nerve, such as the nerve connecting the ear to the brain, cannot transmit twenty thousand impulses per second—in fact, a nerve may have trouble transmitting more than about five hundred or one thousand impulses per second.

Here is how our ear and brain deal with this inability of nerves to transmit such high frequencies: One area of the basilar membrane resonates at twenty thousand vibrations per second. Nerves connected to that area send the brain not twenty thousand nerve impulses per second, but a message that stands for twenty thousand vibrations per second. The brain then translates that message into a pitch sensation.

Incidentally, low notes—like a bass note at one hundred vibrations per second—are handled differently. For a low note, our ear does send one impulse to our brain for each and every vibration of the eardrum—one hundred impulses per second for that bass note.

Further Reading

Roederer, J. G. *Introduction to the Physics and Psychophysics of Music*. 2nd ed. New York: Springer, 1975.

Newton, Tennis, and the Nature of Light

The seventeenth-century English physicist Isaac Newton allowed a beam of sunlight to pass through a triangular glass prism. A rainbow of colors emerged from the prism and fell on the wall opposite the window. Newton wondered why different colors fell on the wall in different places. He wrote, "I began to suspect, whether the rays, after their trajection through the prism, did not move in curve lines, and according to their more or less curvity tend to divers parts of the wall. I remembered that I had often seen a tennis ball, struck with an oblique racket, describe such a curve line" (3078).

A tennis ball struck obliquely or a baseball thrown with a snap of the wrist travels in a curve because it is spinning. As the ball moves through the air, the spin makes air flow faster along one side of the ball than the other. Pressure is lower on the side where the air moves faster. The ball is deflected upward toward the low-pressure side.

Newton toyed with the idea that light might be made of little balls that were somehow made to spin while passing through the prism. Then these balls might follow curved paths from the prism to the wall. Newton soon rejected the curveball idea when he saw that colored light emerged from the prism in straight lines, not curves. Centuries later, physicists learned that particles of light, the so-called photons, bear almost no resemblance to little balls.

But Isaac Newton's attempt to relate light to tennis balls, even though it turned out to be wrong, shows how a first-class thinker searched through all his experience for a picture to help him understand a puzzling phenomenon.

Further Reading

Feynman, Richard P. *QED: The Strange Theory of Light and Matter*. Princeton, NJ: Princeton University Press, 1985.

Newton, Isaac. A letter of Mr. Isaac Newton, Professor of the Mathematicks in the University of Cambridge; containing his new theory about light and colors: sent by the author to the publisher form Cambridge, Febr. 6, 1671/72; in order to be communicated to the R. Society. Philosophical Transactions of the Royal Society 6, no. 80 (December 31 1671): 3075. https://doi.org/10.1098/rstl.1671.0072.

Roll Over, George Washington

Put a quarter on the table with George Washington's head right side up. Hold the quarter down with your finger. Now put another quarter flat on the table, again with Washington right side up, so the edge of the second coin touches the twelve o'clock position on the edge of the first coin. If you roll the second coin halfway around the edge of the first coin, from the twelve o'clock position to the six o'clock position, will George Washington come out right side up or upside down?

As you roll from twelve o'clock to six o'clock, exactly half the circumference of the rolling coin touches the edge of the coin you're holding down. So it seems that the rolling coin goes through half a turn, and Washington should come out upside down. But when you try it, George Washington comes out right side up!

How can that be? If you roll a twenty-five-cent piece along a straight edge, like a ruler, for a distance equal to half its circumference, Washington comes out upside down because the coin rolls half a turn in that situation. But when you roll one twenty-five-cent piece halfway around another, the curvature of the edge of the coin you're holding down with your finger adds another half turn. The two half turns add up. The rolling coin in our demonstration actually goes through one full turn.

And that's how George Washington comes out right side up.

Bibliography

Brown, R. J. *333 Science Tricks and Experiments*. Blue Ridge Summit, PA: Tab, 1981.

Gardner, Martin. *Entertaining Science Experiments with Everyday Objects*. New York: Dover, 1981.

Don't Believe Your Fingers

Most of the illusions we hear about and enjoy playing with are optical illusions. This is a tactile illusion.

To experience this illusion, all you need is a marble or a pea. Cross your fingers, extending the middle finger over the index finger so the two fingertips are next to each other, but reversed from their normal arrangement. Now roll the marble around on a table with your crossed fingertips. Almost immediately you'll probably get the distinct impression that there are two marbles, not just one.

Crossing your fingers makes information travel to your brain through unusual channels. Normally the outside of your middle finger and the inside of your index finger face away from each other. Crossing your fingers brings those two sides together. When the marble gets between your crossed fingertips, it touches areas that normally would be touched only if there were two marbles. Cross your fingers, and one marble feels like two.

This effect has been known long enough to be called Aristotle's illusion. Here are some modern variations.

Try touching your nose instead of a marble. Your fingertips may give you the impression that you have two noses.

Try different fingers: Cross your middle finger over your ring finger. Then lay a pencil over the two crossed fingertips so the barrel of the pencil touches one fingertip and the point touches the other. See if you can tell, just by touch, which way the pencil is pointing. Rocking the pencil gently back and forth may enhance the strangeness and vividness of the sensation.

Further Reading

James, William. "Illusions of the First Type." In *Principles of Psychology*, 86. Vol. 2. New York: H. Holt, 1890.

Opera Singers Cut through the Orchestra

First-class opera singers can make themselves heard distinctly over even a fairly loud orchestra. Acousticians have found that singers accomplish this at least partly by making extra sound at certain moderately high frequencies where the orchestra is not especially loud.

Recall that the sound of a singing voice or a musical instrument is really a complex mixture of vibrations at many different frequencies. Each instrument and each voice has its own peculiar mixture of frequencies that we perceive as its tone color.

Good opera singers learn, by one method or another, to produce a tone that contains an especially large amount of sound energy in the frequency range of two thousand to four thousand vibrations per second. The emphasis on frequencies in the two-thousand- to four-thousand-vibrations-per-second range makes the operatic voice sound very different from the pop singer's voice and from ordinary speech.

The sound of a symphony orchestra, on the other hand, does not have any special emphasis on frequencies between two thousand and four thousand vibrations per second. So opera singers cut through the orchestra by emphasizing frequencies the orchestra does not.

Pop singers, by the way, often don't emphasize those special high frequencies because they want to create a more conversational, speech-like sound than opera singers. But at least one respected book on sound recording recommends that the pop recording engineer electronically amplify the frequencies between two thousand and four thousand vibrations per second on vocal tracks to keep the voices from being buried by the accompanying instruments!

Further Reading

Runstein, Robert E., and David Miles Huber. *Modern Recording Techniques.* Indianapolis, IN: Howard W. Sams, 1986.

Sundberg, Johan. "The Acoustics of the Singing Voice." *Scientific American* 236, no. 3 (March 1977): 82–91.

Curved Space in a Christmas Ornament

The Christmas ornaments we're talking about here are those glass spheres with a mirror-like reflecting surface. Actually, for this observation you don't need a Christmas ornament: any ball-shaped object with a surface that reflects like a mirror will do.

As you look at your reflecting ball, you will see the reflection of your own face at the center. Around your head you will see the whole room—walls, floor, and ceiling. The farther you look from the center of the ball, the more the image is distorted. That's why your nose looks extra-large: the rest of your face is smaller because of the distortion.

As you look more closely, you will see that the ball reflects almost the whole world as seen from the ball's location. The only part left out is the tiny piece of the scene right behind the ball as you view it.

Now suppose someone behind you spreads a square red tablecloth on the floor and measures the edge of it with a ruler. The reflected tablecloth certainly isn't square; the reflection is distorted. But the reflected ruler still shows all four sides to have the same length because the ruler is distorted too.

The distorted reflection in a shiny ball is a model of curved space, one of the more esoteric concepts of mathematics. People living in a curved space may not know they live in a curved space because their rulers and measuring instruments are curved just like everything else.

Further Reading

Minnaert, Marcel. *The Nature of Light and Colour in the Open Air.*
New York: Dover, 1954.

Coriolis Effect

If you were to try to fly a plane in a straight line from Chicago to Atlanta, you would never get there because you would fly to the right of Atlanta. That doesn't seem logical, but it's true. This sideways drifting is called the Coriolis effect, and it's caused by the rotation of the earth. Although those of us who are earthbound don't really have to be concerned about it, airplane pilots have to make adjustments.

To understand how the Coriolis drift affects planes, you need to be aware that if you're standing on the equator, the earth's rotation is carrying you eastward at about one thousand miles per hour. As you move away from the equator, your speed decreases. For example, Boston is traveling at about seven hundred miles per hour.

Now back to our attempt to fly from Chicago to Atlanta in a straight line. As our plane sits in Chicago, it is spinning with the earth, of course, and when it takes off it continues to do so. However, our destination, Atlanta, which is closer to the equator, is moving faster to the east. So if we do not correct for the faster eastward movement, we will end up to the right of Atlanta. (Incidentally, because Atlanta is moving eastward faster than Chicago, if you try to fly northward in a straight line from Atlanta to Chicago, you will still end up to the right of your destination.)

This sideways drifting does not occur just with airplanes; it applies to everything that moves on earth. Were it not for the friction of tires on the road, a car traveling down a highway at sixty miles per hour would be carried off the road to the right at the rate of about fifteen feet per mile.

This chapter has dealt only with how the Coriolis effect applies to navigation, but, among other things, it affects weather and ocean currents as well. But as the saying goes, that's another story.

Further Reading

Ingram, Jay. *The Science of Everyday Life*. New York: Penguin Books, 1989.

McDonald, James E. "The Coriolis Effect." *Scientific American* 186, no. 5 (May 1952): 72–79.

Catch a Falling Dollar

I have here a crisp, straightened-out one-dollar bill. I hold it by a short edge between the thumb and forefinger of my left hand, allowing the bill to dangle flat and vertical. Now I place the thumb and forefinger of my other hand around the bottom edge of the hanging bill, not quite touching it.

I shall demonstrate my quick reaction time by releasing the bill with my left hand and catching it with my right hand before it has time to fall through my fingers. I've done some figuring, and I estimate that a dollar bill takes about a fifth of a second to fall a distance equal to its own length after I release it. I, however, am so quick that I can catch the bill before it has fallen even half its own length. There! I have caught the bill so that my thumb is over George Washington's portrait.

Now you try it. I'll dangle the bill so the bottom edge is between your fingers. Without warning, I drop the bill. You aren't quick enough to catch it. Again and again, you fail.

Actually, I'm cheating. When I drop the bill and catch it myself, I do not demonstrate my own reaction time. My brain issues both the "drop" and "catch" instructions, so I am not really reacting to the motion of the bill. I can make the drop and the catch happen as close together in time as I want. I can even catch the bill before I drop it.

You, however, must see the bill begin to fall before doing anything. Messages must travel from your eyes to your brain to your hand—only then do your fingers close. All that takes time—more time than the dollar bill takes to fall a distance equal to its own length.

Do you think you'll react faster if I use a five-dollar bill instead of a one?

Further Reading

Gardner, Martin. *Entertaining Science Experiments with Everyday Objects*. New York: Dover, 1981.

Illusion in a Coffee Cup

Things are not always as they seem, and this little demonstration will prove it. All you need is a cup of black coffee and an overhead light. A single incandescent bulb works best. Position the coffee cup so the light is reflected in it. Look into the cup from a distance that allows the reflected light to just about fill the cup.

Now move your head quickly and smoothly toward the cup. The light appears to get smaller and farther away! The change is dramatic. The light seems to shrink to a quarter or a fifth of its original size and to move away ten times more than you moved.

This is obviously an illusion since you know the light hasn't changed size or moved. However, as you move closer to the cup, the cup fills more of your field of vision than the light, so the cup seems larger and the light smaller.

In trying to interpret the information being sent to it, your brain correlates smaller with farther away, and you perceive that the light has moved. Your brain cannot make an adjustment, either; you can do this over and over again and the result will be the same.

Further Reading

Ingram, Jay. *The Science of Everyday Life*. New York: Penguin Books, 1989.

Senders, John. "The Coffee Cup Illusion." *American Journal of Psychology* 79, no. 1 (March 1966): 143–45.

Why Do We Put Cut Flowers in Water?

Water keeps cut flowers and other plants crisp because of one of the most important and all-pervasive natural processes operating on the face of planet Earth: osmosis. Osmosis is the process in which liquid water tends to move toward regions with a higher concentration of dissolved substances. The dissolved substances might be minerals, sugars, anything—water will usually move to where there's more dissolved material.

Each cell of a plant has a sort of skin—a membrane. Water can pass through the membrane easily, but other materials can't. Each plant cell maintains a relatively high internal concentration of dissolved materials. Water therefore tends to move into the cell.

As long as the concentration of dissolved substances is higher inside the cell than outside, water will usually push its way in. The water pressure that builds up inside the cell is what gives a healthy plant its crisp texture. Often that pressure will make cells expand; that's one way plants grow. And from this you can see why plants wilt if they don't get enough water: the cells lose internal water pressure.

So plants in general, and cut flowers in particular, stay crisp because dissolved materials in effect draw water through the cell membranes into the plant cells by the process of osmosis. There are other ways living things move water from one place to another, but osmosis is one of the most important.

Further Reading

Curtis, Helena. *Biology.* 4th ed. New York: Worth, 1983.

Salisbury, Frank B., and Cleon W. Ross. *Plant Physiology.* 3rd ed. Belmont, CA: Wadsworth, 1985.

Knuckle Cracking

How many of us have derived impish delight from annoying our mothers by cracking our knuckles? It's not something I did a lot, but I do recall my mother trying to discourage me by telling me it would cause my joints to swell. But then again, she also told me my hair would fall out if l wore my hat indoors.

While the annoyance factor related to knuckle cracking is indisputable, the actual cause of the noise was a puzzle until the early 1970s and has been hotly debated since. The most common explanations for the noises were bones snapping against each other or tendons moving over bony projections in the joints. It kind of gives you goose bumps just thinking about it, doesn't it? Well, this chapter is going to tell you what really makes the sound of cracking knuckles. The sounds come from tiny explosions. Now, if the thought of your joints exploding doesn't make you feel any better, let me clarify. The sounds are not your joints exploding, but the popping of gas bubbles in the lubricating fluid that fills the joint. In order to crack your knuckles, you stretch the joints, causing an increase in the space between the finger bones. This increase in space reduces the pressure on the fluid that lubricates the joints. This reduction in pressure causes tiny gas bubbles to form in the fluid. As the pressure continues to go down, the bubbles burst, making the popping noise you hear.

After the bubbles burst, the gas does not escape the area, but is reabsorbed into the fluid as the joint returns to its original position. It takes about fifteen minutes for the fluid to be reabsorbed, and that's why once you crack your knuckles, you can't do it again for a while.

So, even though the evidence indicates you will not suffer dis-figurement from cracking your knuckles, for the sake of mothers past, present, and future, if you feel the need to pop your knuckles, don't do it around your mother—or anyone else's.

Further Reading

Kluger, Jeffrey. "Why Does Cracking Your Knuckles Make So Much Noise? Science Finally Has an Answer." *Time*, March 29, 2018. https://time.com/5220275/knuckles-crack -science-why-reason/.

Walker, Jearl. *The Flying Circus of Physics with Answers*. New York: John Wiley and Sons, 1977.

Life without Zero

Zero is one of humanity's greatest inventions, a symbol that stands for nothing in a very definite way.

The ancient Greeks and Egyptians had no zero. They used completely different symbols for nine, ninety, nine hundred, and so on. This system had a couple of big disadvantages. First, it had symbols only for numbers people had already thought of. If you had wanted to talk about, say, nine hundred billion, you would have had to invent a symbol for it. The old Greek and Egyptian systems also make arithmetic hard. Without zero, multiplying three times ninety is a whole different problem from multiplying three times nine.

The first known zero symbols appear in Babylonian clay tablets of about 500 BC; there, the zero was used to clarify the symbols for large numbers.

The idea that zero can be treated in arithmetic problems as a number like any other number came from Brahmagupta, a Hindu astronomer of the seventh century AD. He was the first to write down the rules for arithmetic with zeros. Western civilization didn't adopt arithmetic with zeros until about seven hundred years later, based on the work of the thirteenth-century Italian mathematician Leonardo Fibonacci.

Thanks to zero, we have to learn multiplication tables only up to ten times ten. Thanks to zero, we can punch any number into our calculators using just ten keys. And if we want to imagine some gigantic number, we can do it easily—just by adding more zeros.

Further Reading

Dictionary of Scientific Biography, s.v. "Brahmagupta." New York: Charles Scribner's Sons, 1970.

McGraw-Hill Encyclopedia of Science and Technology, s.v. "Zero." New York: McGraw-Hill, 1987.

Neugebauer, Otto. *The Exact Sciences in Antiquity*. New York: Dover, 1968.

For This You Need a Doctor?

Chicken soup is a time-honored treatment for fevers and coughs and has been recommended since ancient times by physicians and grandmothers. At least one modern scientist daring to investigate the validity of this ancient advice found it to be correct.

Stephen Rennard at the University of Nebraska Medical Center wanted to see if there was a scientific basis for the chicken soup remedy. He knew that when the body is invaded by a virus or harmful bacteria, it sends certain white blood cells to release enzymes to help fight the infection. Unfortunately, the enzymes that fight the infection also irritate our tissues and give us the sore throats that so often come with the flu.

Rennard devised an experiment in which he could observe the movement of white blood cells toward bacteria, similar to the movement that takes place in the human body. He hoped to test what effect, if any, chicken soup had on this movement.

Using a recipe from his wife's grandmother, he added the various ingredients of the soup in stages to a container that held white blood cells and bacteria. Plain water had no effect on the movement of the white cells toward the bacteria, but when he added the vegetables, the cells moved distinctly slower. Unfortunately, the vegetable broth also killed some of the body's white cells.

However, when he added chicken to the vegetable soup recipe, no harm was done to the cells. Evidently the chicken counteracted the toxic effects of the vegetables. The cells' movement was still slowed, so Rennard assumed that in a human body the slowing down could reduce the number of enzymes at the site of infection and probably reduce the inflammation and some of the discomfort. Interestingly, the slowed movement did not seem to affect the cells' infection-fighting abilities. Also, this bit of research showed that although a remedy might be traditional, that doesn't mean there isn't a scientific basis for it.

Further Reading

Cerino, Vicky. "Got a Cold or Flu? UNMC Researcher Says Try Chicken Soup." University of Nebraska Medical Center Newsroom, October 4, 2012. https://www.unmc.edu/news .cfm?match=9973.

Rennard, Barbara O., Ronal F. Ertl, Gail L. Gossman, Richard A. Robbins, and Stephen I Rennard. "Chicken Soup Inhibits Neutrophil Chemotaxis *In Vitro*." *Chest* 118, no 4. (October). 1150–57. https://www.unmc.edu/publicrelations /media/press-kits/chicken-soup/chickensouppublished study2000.pdf.

Two-Point Threshold

Here's a fun experiment that tells you something about your nervous system. You can do it on yourself, but it works best if you do it on someone else and vice versa. All you need is two pointed objects, such as pencils.

Have your friend close her eyes, and tell her you're going to touch the inside of her forearm with the two pencil points simultaneously about eight inches apart. Then you're going to lift the points and bring them down again and again, each time moving the points closer together. She's to tell you when you're touching her with only one pencil. It's surprising that when she thinks she's being touched by only one point, both are still touching her arm, and they are one to two inches apart!

Now do the same thing on the pad at the end of her index finger, but start with the points about an inch apart. This time when she thinks she feels only one point, the two will be about an eighth of an inch apart.

Why does she feel only one point when she's being touched by two? And why are the distances where she feels only one point as much as two inches apart on the forearm, but only an eighth of an inch apart on the fingertip? It's because some areas of the body, such as the fingertips, have more nerves going to them. And the areas of the brain that receive information from these sensitive areas of the body have a greater density of nerves. Since more nerves are present to detect sensations, these areas are more discriminating.

This is an interesting experiment, but there is a useful side to the information. This psychological phenomenon is known as the two-point threshold, and it can be used to test for nerve damage.

Also, Braille dots were designed to be farther apart than the two-point threshold.

Further Reading

Geldard, Frank A. *Fundamentals of Psychology*. New York: John Wiley and Sons, 1962.

Schiffman, Harvey R. *Sensation and Perception: An Integrated Approach*. 2nd ed. New York: John Wiley and Sons, 1982.

A Mirror Riddle

Exactly what is the difference between the appearance of a real object and its reflection in a mirror?

Obviously, a mirror reverses the image of an object in some way. For instance, when you look into a bathroom mirror, you see an image with left and right switched. If you hold up a toothpaste tube, the letters on the reflected tube are backward. Evidently, reflection in a bathroom mirror reverses left and right.

But not all reflections reverse left and right. Some switch top and bottom. Think of trees on the far side of a lake and how they're reflected in the water. The reflected trees have top and bottom switched, but left and right remain the same for the reflected trees and the real trees.

It seems like a contradiction: A bathroom mirror switches left and right, and the surface of a lake switches top and bottom. How can that be? (The explanation, by the way, has nothing to do with the difference between glass and water. If you put a mirror flat on a table, the reflections will be oriented the same way as in the surface of a lake.) Is there some precise way of describing the essential difference between a real object and its mirror image— some rule that will work in every situation?

Here's a hint: Look at the reflection of a clock in a mirror— the old-fashioned kind of clock, with hour and minute hands. A second hand is even better. What's the essential difference between the real clock and its reflection?

Think about what you saw when you looked at the reflection of the clock. While the hands on a real clock run forward, or clockwise, the hands on the reflected clock run backward, or counterclockwise. Therein lies the answer to our riddle: a mirror switches the clockwise and counterclockwise directions. Whether the reflection is right side up, upside down, or sideways simply depends on where you put the mirror. But the essential difference between reality and reflection is the reversal of clockwise and counterclockwise.

Look at the letter *p*, as in *toothpaste*, on your toothpaste tube. The shortest trip from the top of that letter *p* around the loop is clockwise on the real tube and counterclockwise in the upside-down reflection in the water. On the face of your clock, the shortest trip from the twelve to the three is clockwise in reality and counterclockwise in the mirror.

A mirror switches clockwise and counterclockwise. That surprisingly subtle rule is the only one that always works, regardless of the position of the mirror, or the person looking, for any reflected image.

Sort Nuts by Shaking Can

If you shake a can of mixed nuts for a few seconds, the largest nuts come to the top. The spaces between the nuts are not big enough for small nuts to fall through, but the small nuts end up on the bottom and the large ones on top anyway. Shaking creates momentary gaps in the mixture; small gaps occur more often than large ones.

Then the nuts are sorted by gravity. As the can shakes, the large nuts frequently move aside far enough to allow a small nut to fall into the space beneath. This happens much more often than the reverse process, in which several small nuts happen to make a gap that one large nut can fall into.

Every time a large nut moves far enough to allow a smaller nut to fall into a gap beneath it, that large nut ends up resting on top of the smaller nut. Over the course of several seconds of shaking, the large nuts slowly move up.

This is not only an amusing kitchen observation, but something with practical usefulness. In many parts of the world, construction workers separate gravel from sand by shaking the container. Coarse gravel comes up to the top. Manufacturers can exploit the tendency of large particles to move up when they need to make a mixture of particles of different sizes in pharmaceuticals, glass making, and paint making. They can put large particles into a container first and then add smaller particles on top. Shaking the container for the right amount of time causes the large particles to move up until they're evenly distributed through the mixture.

Further Reading

Rosato, Anthony, Katherine J. Strandburg, Friedrich Prinz, and Robert H. Swendsen. "Why the Brazil Nuts Are on Top: Size Segregation of Particulate Matter by Shaking." *Physical Review Letters* 58, no. 10 (March 9, 1987): 1038–40.

Prigo, Robert B. "Liquid Beans." *The Physics Teacher* 26, no. 2 (1988). https://doi.org/10.1119/1.2342442.

Science and the Citizen: Physical Sciences: Nuts and Bolts. *Scientific American* 256, no. 5 (May 1987): 58D.

Why a Rubber Band Snaps Back

Stretch a rubber band; it becomes long and thin. Let it go; it snaps back to its original short, fat shape, ready to be stretched again.

Rubber has this useful property for two reasons. First, rubber molecules have a peculiar structure and arrangement; second, those molecules are always moving around because the rubber is warmer than a temperature of absolute zero. In any material warmer than absolute zero, the molecules are always moving around in a tiny random, jiggling motion.

Rubber is made of molecules shaped like strands of spaghetti. If you stretch a rubber band, you pull those spaghetti-shaped molecules into a more or less straight line. But the molecules are still moving around. They shake from side to side and bump into each other. Because of that motion, the molecules tend to spread out sideways, and they also "unstraighten"—curl up, kink, and tangle. That makes them pull inward on the ends of the rubber band. The stretched rubber tries, so to speak, to become short, thick, and flabby so the molecules will have more room to move around sideways. The rubber band snaps back.

This also explains another property of rubber: Unlike most materials, it shrinks when you heat it and expands when you cool it. In a warm rubber band, the molecules move faster, tend to shake sideways more, and therefore pull harder at the ends than in a cold rubber band. A rubber band will squeeze a package harder if it's been in the sun than if it's been in the freezer.

Further Reading

Feynman, Richard P. *The Feynman Lectures on Physics*. Vol. 1. Reading, MA: Addison-Wesley, 1964.

Wall, Frederick T. "Statistical Thermodynamics of Rubber." In *Chemical Thermodynamics*. 2nd ed. San Francisco: W. H. Freeman, 1965.

Some Like It Hot

Most of us have jumped into a swimming pool and felt the shock of the cold water, only to have it feel just fine after a minute or so. Or we've stepped into a nice warm shower and after a minute reached over and turned up the heat a little because it no longer felt warm enough. Those who like hot showers might turn up the heat a couple of times before it suits them. This ability of the body to adjust to temperatures is called thermal adaptation. When you adapt to a temperature, it means it doesn't feel cold or hot, but neutral.

A simple experiment can clearly demonstrate thermal adaptation. Get three bowls large enough to put your hands in. Put cold water in one—you might add a few ice cubes to make the water cold. In the second put water that is hot, but not too hot to put your hands in. And in the third bowl put water that is warm, about ninety degrees.

Put one hand in the bowl with the cold water and the other in the hot water for about a minute. Now put both hands in the bowl of warm water. The warm water will feel very cool to the hand that originally was in the hot water and warm to the hand that was in the cold water.

Fortunately thermal adaptation has its limits, because otherwise we might get burned or frozen if at some point the warning sensation of extreme heat or cold didn't override adaptation. Scientists do not completely understand the process of thermal adaptation, but they've known about it for a long time. The first report of the experiment with the three bowls of water was given by the philosopher John Locke in 1690.

Further Reading

Geldard, Frank A. *Fundamentals of Psychology.* New York: John Wiley and Sons, 1962.

Schiffman, Harvey R. *Sensation and Perception: An Integrated Approach.* 2nd ed. New York: John Wiley and Sons, 1982.

Breaking the Tension

Carefully fill a glass with water until the surface of the water is exactly level with the rim of the glass. You may want to use a second glass to add the last of the water, rather than trying to do it at the faucet. Now gently drop a quarter into the glass. The water surface will bulge upward slightly, but water will not run over the rim.

Molecules of water attract each other. That attraction makes a film under tension at the surface of the water. Even though the water is bulging upward, surface tension keeps it from spilling.

Now the game is this: How much change can you drop into the water before water spills over the rim of the glass? Every coin displaces an amount of water equal to its own volume. In other words, the bulge at the top of the water has the same volume as all the coins added to the glass. The film at the surface of the bulging water caused by attraction of water molecules holds the water like a bag. How much water can that bag hold before it breaks?

Possibilities for elaboration suggest themselves. You could agree that the next-to-last person to add a coin before the water spills wins all the change in the glass.

To make the game last longer, use paper clips instead of coins. As before, carefully fill the glass with water until the water surface is exactly level with the glass brim. How many paper clips can you drop in before the surface tension breaks and water spills over the edge? Some people may guess that about ten paper clips will do it. But remember, a paper clip is just a piece of bent wire, which displaces only a tiny volume of water. You may be able to get over a hundred paper clips into the glass before the water spills.

Further Reading

Gardner, Martin. *Entertaining Science Experiments with Everyday Objects*. New York: Dover, 1981.

Why 5,280 Feet?

Our word *mile* comes from the Latin *mille*, which referred to the Roman mile. The Roman mile had military origins since it was the equivalent of one thousand double paces of marching soldiers. The soldiers' double paces were about five feet, so the Roman mile was about five thousand feet.

Since we got our measurement system of inches, feet, yards, and miles from the British, what does the Roman mile have to do with our mile? Well, Britain was part of the Roman Empire from the first to the fifth centuries AD, so when the British began to standardize their measuring system, there was a Roman influence.

Even before the British started keeping written records of landholdings, farmers laid out their fields in plowed furrows that were consistently the equivalent of a modern 660 feet long. This distance became a standard part of their measurements.

Over time, by slurring the words, this "furrow-long" distance became *furlong*, a unit now used almost exclusively in horse racing.

The British eventually used the Roman mile as a model in their measurement system, but they didn't want to give up their furlong. The Roman mile was about seven and a half furlongs, and when the British adopted it, they lengthened the Roman mile to eight furlongs, which equals 5,280 feet.

Further Reading

Feldman, David. *Imponderables: The Solution to the Mysteries of Everyday Life.* New York: Quill, 1987.

Balance a Yardstick without Looking

A yardstick is thirty-six inches long, so you know the exact center is at the eighteen-inch mark. But how could you find the center blindfolded?

You might guess something complicated, like using the distance between two knuckles to represent one inch and measuring with that. But there's an easier way. Hold your hands in front of you, with your index fingers pointing forward. Rest the yardstick across your fingers and slowly bring your hands together under the stick. Your fingers will meet under the eighteen-inch mark. You're actually finding the center of gravity of the stick, the point where it balances. We're assuming that your yardstick is made of wood or another material that's uniform from end to end, with no metal attachments or big holes that would cause it to balance somewhere other than the center. If that assumption is correct, the yardstick will balance at the eighteen-inch mark.

As you bring your hands together, you will feel the stick sliding first over one finger, then over the other. Whenever your left hand gets closer to the center of gravity than your right hand, the yardstick will press down harder on your left index finger. More pressure causes more friction, so the left hand will stop sliding. All the sliding will then happen over your right index finger until your right hand gets closer to the center and reverses the situation. In other words, if one hand gets ahead, friction stops it from sliding until the other hand catches up.

All this happens with no special effort on your part. Just rest the yardstick across your index fingers and slowly, steadily, bring your hands together. Your hands will meet at the eighteen-inch mark, the center of gravity of the yardstick.

Further Reading

Gardner, Martin. *Entertaining Science Experiments with Everyday Objects*. New York: Dover, 1981.

Steinhaus, H. *Mathematical Snapshots*. New York: Oxford l University Press, 1950.

Heat Lightning

One of the more mysterious pleasures of a warm summer evening is the spectacle of lightning from distant thunderstorms flickering silently on the horizon while stars shine overhead. People usually call it heat lightning.

Lightning is easy to see at great distances, especially when it illuminates high, thin clouds visible for many miles. But thunder usually doesn't carry more than about ten or fifteen miles from the storm because turbulent air around a storm acts as a damper on sound waves.

Another reason thunder doesn't carry very far has to do with differences in temperature between air at ground level and higher up. Early on a summer evening, the ground is still warm from afternoon sunshine, so the air at ground level is also warm. A few thousand feet up, the air is cooler. This temperature difference bends the sound of thunder upward.

Here's how it happens. We can think of a sound wave as an invisible wall of slightly compressed air traveling at the speed of sound. Sound travels slightly faster in warm air than in cool air. So the part of that invisible wall down in the warm air travels a little faster than the part in the cool air higher up. The bottom of this invisible wall gets ahead of the top as it travels. The invisible wall of sound bends upward as it goes.

The common term *heat lightning* actually describes an essential feature of the situation. Because of a layer of warm air near the ground, the sound of thunder is bent upward, into the night sky. The result is that you see lightning on the horizon but you don't hear thunder.

Further Reading

Ahrens, C. Donald. *Meteorology Today*. 2nd ed. St. Paul, MN: West, 1985.

The Moon Illusion

For thousands of years, people have been noticing that the moon looks much bigger when it's very low in the sky than it does when it's high overhead. This effect has come to be called the moon illusion.

It's not a physical effect; the atmosphere does not magnify the image of the moon. The atmosphere may cause the moon to appear flattened or colored, but not magnified. Photographs show that the moon's image is really the same size no matter how high or low it is in the sky. But people almost always judge the image to be larger when the moon is low.

The real cause of the moon illusion seems to be the juxtaposition of the moon and features on the distant horizon. Our perception of the distance to the horizon influences our judgment of how big the moon is. Professional psychologists have done experiments that indicate this, but you can investigate the moon illusion yourself.

Cut a hole the size of a quarter in a big piece of cardboard. Hold the cardboard at arm's length, not next to your eye. Look at the rising or setting moon through the hole. That cardboard screen blocks your view of the landscape near the moon. Does the moon look smaller with the cardboard than it does without?

Another experiment involves looking at the whole scene upside down by bending over and looking between your legs. For most people, looking at a scene upside down weakens the impression that the horizon is far away. Does that weakened impression of distance dispel the moon illusion for you?

Looking at the rising or setting moon through a hole in a cardboard screen and looking at it upside down are two simple tricks that may dispel the moon illusion. Try them to get a hint of the mystery and complexity of our ability to judge distance and size.

Further Reading

Kaufman, Lloyd, and Irvin Rock. "The Moon Illusion." *Scientific American* 207, no. 1 (July 1962): 120–31.

Warmth from a Cold Lamp

Find a lamp that has not been on for several hours so that the light bulb is stone cold. Touch the bulb with your fingers. It should feel just about as warm or cool as any other glass object in the room.

Now, with your hand near the bulb, but not touching it, switch the lamp on for five seconds, then switch it off. During the five seconds the lamp is on, you will feel what seems to be heat coming from the bulb. Touch the bulb again immediately after you switch the lamp off; the bulb will feel almost as cool as it did before. It seems paradoxical—a light bulb that's not hot can make your hand feel warm.

What this experiment really demonstrates is that light and heat are not the same thing. When you turn on the bulb, what hits your hand is light—lots of visible light, and even more invisible infrared light. Light of any kind is really energy traveling through space. When light, visible or not, strikes matter—your hand, for instance—that energy can be converted to other forms. Here, the energy of light is converted to tiny random motions of the atoms in your skin: in other words, heat.

Heat can also be communicated directly by matter. If your hand is above the light bulb, you will soon receive additional heat from air, warmed by the bulb, rising to your hand. But you will feel warmth as a result of the light even if your hand is in the cool air below the bulb.

The sun warms the earth by the same process: sunlight—visible and invisible—travels through ninety-three million miles of airless space, changing to heat only when it strikes the earth or some other object, such as a spacecraft. It's true that the sun's surface is hot—like the filament inside the light bulb—but that heat is not directly communicated to faraway objects. The *light* from the sun is what warms the earth.

Further Reading

Miller, Julius Sumner. "Further Enchanting Things to Think About." *The Physics Teacher* 17, no. 6 (September 1979). https://doi.org/10.1119/1.2340275.

DON GLASS is the producer of *A Moment of Science* which originates from Indiana University's public radio station WFIU. He has been associated with the program since its inception in 1988.